W9-BVV-992

THE ABYSS OF TIME

The Abyss of Time

Changing Conceptions of the Earth's Antiquity after the Sixteenth Century

Claude C. Albritton, Jr.

AN ERNEST SCOTT BOOK

JEREMY P. TARCHER, INC.
Los Angeles
Distributed by St. Martin's Press
New York

Library of Congress Cataloging in Publication Data

Albritton, Claude C., 1913-
The abyss of time.

Reprint. Originally published: San Francisco, CA : Freeman, Cooper, c1980.
Bibliography: p.
Includes indexes.
1. Geological time. 2. Geology—History. I. Title.
QE508.A47 1986 551.7 86-1755
ISBN 0-87477-389-X

Jeremy P. Tarcher, Inc.
9110 Sunset Blvd.
Los Angeles, CA 90069

Manufactured in the United States of America
10 9 8 7 6 5 4 3 2 1

First Edition

for Jane Christman Albritton

Contents

ILLUSTRATIONS

To the Reader

I am confident that someday the concept of geological time will be acclaimed as one of the more wonderful contributions from natural science to general thought.

Geological time measures the age of the earth, from its beginnings as a planet to the present. By this definition, the *prehistory* of the archaeologists and the *historical time* of the historians are respectively the next-to-last and last intervals in the continuum of geological time.

A debate about the earth's age has been going on for centuries, and though a precise figure in years remains to be determined, an order of magnitude now seems firmly established. The earth originated several thousands of millions of years ago. In that perspective, the intervals of time covered by prehistory and history are seen to be astonishingly brief.

The concept of the vast extension of our world in time, one which grew from investigations seeded in the seventeenth century, has surely been no less revolutionary than the concept of vast stellar space emerging after Copernican astronomy appeared during the century preceding. However, outside a relatively small community, which includes most practitioners, historians, and philosophers of science, many conservationalists and writers of science fiction, and a growing number of artists and poets, the idea of geological time as often as not has been greeted with indifference, skepticism, or downright hostility.

Charles Lamb probably delivered the majority opinion for the literate minority when he declared: "Nothing puzzles me more than time and space; and yet nothing troubles me less, as I never think about them." And yet Lamb must have been thinking about time and space when he wrote that sentence, and he hit on the right words too when he chose "puzzles" and "troubles."

Assuredly time is one of the more puzzling of human inventions. Time can take various shapes according to our choice. Time can be conceived as circular or linear, continuous or interrupted, finite or infinite. The Stoics thought of time as circular and continuous, as though historical events were arranged in sequential order around a great celestial wheel, which in its turnings causes history to repeat itself world without end. In the Stoic view, events are like the succession of sounds on a turning phonograph record with a "hung" needle, or like the markings on a roulette wheel which in its spinning causes the same succession of numbers to move again and again past some point on the perimeter. On the other hand, the ancient Mayas seem to have conceived time as linear and without repetition of events, but with periodic pauses in its flow when nothing happens. Modern historians and practitioners of historical science assume that time is linear and continuous, without cyclical recurrence of events. Indeed, if time were not puzzling, there would not be an International Society for the Study of Time, which has published three thick volumes of proceedings since 1971.

The concept of time can also be intimidating, to institutions as well as to individuals. Father Time—that animated skeleton armed with a scythe—will mow each of us down soon or late. Most metaphors for time are discouraging: time is a poison, or an acid that dissolves life, or a thief in the night, or a vast and lifeless desert, or a river that sweeps us away in its flood. The comparison of time with a fearsome abyss is common in writings from the seventeen hundreds, though this conceit goes back much earlier, as we know from lines in *The Tempest* recording Prospero's query to Miranda about "the dark backward and abysm of time." And, as standing on the lip of a canyon and looking downward may arouse that species of space-related fear called acrophobia, so pondering the conceptual abyss of geological time may bring on a bad case of chronophobia.

To my knowledge, there are no formal organizations of acrophobiacs. But the chronophobiacs are well organized, and vocal. Their organizations go by different names, of which the Creation Research Society is one. Members of the CRS persistently seek to promote teaching in the public schools of their creationist doctrine, as an "alternative" to evolutionary biology. Though evolution is their avowed target, they must necessarily confront the formidable array of evidence that fossils provide in support of evolutionary theory. And to explain away that evidence they insist that the earth cannot be nearly so ancient as the geologists claim. In the last chapter I call attention to some of the interesting convolutions in their arguments.

I must add that the purpose of this writing is not to air my disagreements with members of the Creation Research Society. I mention them only to make the point that modern scientific estimates for the duration of geological time are not universally accepted, even by persons who hold academic degrees.

To trace successive changes in perceptions of the earth's antiquity is the purpose of this book. I do not attempt to cover the earliest speculations on the subject, but begin with developments during the mid sixteen-hundreds—the time when historical geology began to branch from the trunk of natural philosophy. Even with this limitation, the subject is still too big to be covered in a book of this length. So I have concentrated on episodes foreshadowing or signalling changes in views on geological time, and on some of the persons who played prominent roles in bringing those changes about.

None of the chapters has been published previously, but the contents of most have been presented orally by me at various times and places. The second and third chapters are revised and shortened versions of lectures delivered in 1970 at the University of Pennsylvania, in the course of my appointment there as Rosenbach Fellow in Bibliography. The substance of these and succeeding chapters has been presented less formally on other occasions to a selected group of high school students attending Saturday seminars sponsored by the National Science Foundation, to undergraduate and graduate students in classes at Southern Methodist University, and once to a group of university professors and their spouses at a Danforth Conference in Estes Park. To my satisfaction, I have found that sampling the history of thought relating to the earth's age seems to have been an engaging exercise for many persons with reflective turns of mind, regardless of their age or prior exposure to geological principles. I only hope I've chosen the right samples here, from the many available in the literature of more than three centuries.

The idea of summarizing these lectures in book form belongs to W. H. Freeman. That the work of writing has spread over a decade was due to my desire to visit and examine at first hand certain classic localities abroad. Accordingly, with the aid of a grant from funds provided by the Danforth Foundation and of leaves extended by my university, I have tracked Steno over Tuscany, Lyell around Etna, Scrope across the Auvergne, Hutton to Siccar Point, and many others to places where historical geology finds its roots.

For their help in providing essential source materials, I am indebted to many members of the library staff at Southern Methodist University, especially to Deverett D. Bickston, Mattie Sue Mounce, and James G. Stephens of the Science/Engineering Library, and to James W. Phillips, who of all doctors of philosophy is the one best acquainted with the rare books on geology in the DeGolyer Western Collection.

Wherever possible I have let those writers who have influenced our ideas concerning geological time speak for themselves, through numerous quotations selected from their works. Publishers of writings still protected by copyright have been unanimously generous in permitting the use of quotations from their books and journals. Permissions granted by the American Association for the Advancement of Science, Random House, St. Martin's Press, Taylor & Francis Ltd., University of Chicago Press, and University of Wisconsin Press are referenced at appropriate places in the text and are gratefully acknowledged here.

In order to obtain copy for the illustrations, I had to appeal to the good offices of many individuals and institutions. Professors Donald B. McIntyre of Pomona College and Harry B. Whittington of the University of Cambridge together provided portraits of Hutton, Playfair, and Sedgwick. Dr. Whittington volunteered the additional favor of placing me in correspondence with Ms. Rosemary Evans, Archivist to the Geological Society of London, who graciously sent me prints from freshly prepared negatives portraying Scrope, Lyell, Murchison, and William Smith. Permissions to reproduce portraits of Rutherford and Chamberlin were granted respectively by the Royal Society of London and the National Academy of Sciences. With the concurrence of Professor Albert V. Carozzi, the University of Illinois Press gave permission to reproduce his drawings illustrating de Maillet's theory of the earth, together with a portrait of de Maillet. The two illustrations of geological subjects from the lost drawings of James Hutton are reprinted with the permission of Sir John Clerk of Penicuik; and the remaining illustrations were assembled from local sources with the expert aid of Dr. Phillips.

Ms. Eleanor Swank has not only typed most of the manuscript but also served as a critic for clarity of expression. Professor Konrad B. Krauskopf has reviewed the text for scientific accuracy. Mrs. W. H. Freeman has tended to the onerous task of copy-editing, and from time to time has given me much-needed encouragement to bring this work to completion. For any residual and undetected flaws, I am of course solely responsible.

CLAUDE ALBRITTON
Dallas, Texas

Summer, 1980

We regret the following error in the first printing:

Page 125. Charles Lapworth was British, not a Scot.

THE ABYSS OF TIME

Time is but the stream I go a-fishing in. I drink at it; but while I drink I see the sandy bottom and detect how shallow it is. Its thin current slides away, but eternity remains. I would drink deeper; fish in the sky, whose bottom is pebbly with stars.

THOREAU

CHAPTER ONE

Introduction

The figures now considered by scientists as providing reasonable estimates of the earth's age are more than seven hundred thousand times greater than those in vogue three centuries ago. This expansion has not been upward on a uniform slope. During the seventeenth century the age of the earth was reckoned in the thousands of years. Estimates in the millions and even thousands of millions followed in the eighteenth century. During the late nineteenth century there was a general shrinkage to limits that fell as low as a few tens of millions. Now we think in terms of thousands of millions. These fluctuations have not been capricious, but have been prompted by increase in knowledge concerning the nature of matter in general and of that particular matter available for inspection in the outermost parts of the solid earth.

Some of the earlier estimates were based upon interpretations of Scripture. During the fourth century of the Christian era, Eusebius of Caesarea devised a chronicle based on Jewish historical traditions reaching back in time to the birth of Abraham. Eusebius Hieronymus (Saint Jerome) extended the chronicle back to Adam. Efforts to derive a history of the earth and its inhabitants from Biblical sources culminated in the seventeenth century with the work of James Ussher, Irish scholar and theologian. Ussher's chronology placed the date of creation in the year 4004 B.C., and this figure has since been imprinted in many editions of the Bible.

The scientific revolution of the seventeenth century did not produce developments in geology so spectacular as those in mathematics, astronomy, physics, and chemistry. Nevertheless, two ideas which would later have great influence on historical geology had emerged before 1700. Fossils were coming to be accepted as remains of organisms rather than sports of nature. And the layered rocks containing these fossils were recognized as having accumulated layer by layer. Thus the order in which the layers are stacked is chronologic. The *principle of superposition of strata*, which embodies this concept, states that in a sequence of strata, as originally laid down on the surface of the solid earth, any stratum is younger than the one it rests upon and is older than the one that rests upon it.

The first of two time-explosions came during the 1700's. Estimates of the earth's age based on experiments with earth models yielded figures up to three million years. Speculations based upon a supposed secular shrinking of the world ocean led to a figure of two thousand million years. A leading scientist of the century, on viewing the evidence for repeated elevation and wearing down of the continents, concluded that geological time must be inconceivably long—virtually infinite in duration—in order to accommodate events of such magnitude.

Studies in France and England conducted between 1779 and 1815 established that within thick sequences of fossiliferous strata different kinds of fossils succeed one another in a definite order. This *principle of faunal succession*, combined with the principle of superposition, made possible the classification and chronologic ordering of fossiliferous rocks exposed over continental Europe and the British Isles. Individual strata were grouped into formations, and formations were combined into higher categories leading up to the geologic systems that we recognize today. The laborious task of classifying and identifying the chronologic order in which the systems are arrayed was essentially accomplished by 1845. At that time, however, the chronicle provided by the systems was only relative. One could say that a given system is younger than the one it overlies and older than the system next above, but not by how many years.

Until after the mid-nineteenth century, geologists were prone to claim almost any amount of time needed to account for the myriad events recorded in the rocks. Assuming an indefinitely large allowance of time, Darwin constructed his theory of organic evolution on the basis of the slow-working process of natural selection.

Soon after Darwin's *Origin of Species* appeared in 1859, however, there was a sharp recession in the amount of time thought to be allowable for the history of the earth and the evolution of life. On the assumption that the earth is an inert mass that was originally molten, Lord Kelvin calculated the time that would be required for cooling to its present condition. The figures he proposed varied throughout the years when he addressed this problem; but by the end of the century he was convinced that the earth could not have supported life for more than 20 to 25 million years.

Geologists reacted to Kelvin's abbreviated time scale by devising numerical scales of their own. One method of arriving at numbers was to divide the maximum thickness of the geological column by the average yearly rate at which sediments are accumulating at present. Another "hourglass" approach used the ratio between the total amount of sodium in sea water and the amounts of that element annually added at present. In either instance one could arrive at nearly any figure he pleased, by juggling values of the several variables within whatever seemed permissible limits. Not surprisingly, many of the quotients fell out around a hundred million, an age Kelvin had suggested during his more generous years.

A second time-explosion was triggered in 1896 by the discovery of radioactivity. The facts that naturally radioactive substances are widely disseminated in the outer parts of the earth and that radioactive decay entails release of heat put an end to Kelvin's time scale. Beyond that, ratios between parent and end products in certain lineages among radioactive series yield ages in years for the substances that contain them. Radiometric dating is mainly responsible for the calibration of the relative scale of time worked out by the geologists. Results to date indicate that the earth is more than 3,700 million years old.

This bare sketch of the ups and downs of geologic time now completed, we can turn to the whys and the wherefores and the interesting persons responsible for these changes in our perspective on time.

CHAPTER TWO

Malta's Cradle

> Above the plains of Italy, where flocks of birds are flying today, fishes were once moving in large shoals.
>
> LEONARDO DA VINCI

In the autumn of 1666 some men fishing in the Ligurian Sea caught a great white shark and dragged it ashore near Leghorn. The landing of a white shark was a notable event then, as it is today, for of all fish this is considered the most dangerous to man. Word of the catch spread to the Medici Court in Florence. By order of the Grand Duke Ferdinand II the shark's head was cut off and brought for examination to one Nils Stensen, M.D., a twenty-nine-year-old Dane who had latinized his name to Nicolaus Stenonis (now usually anglicized to Steno). The chain of events which led to the conjunction of an extraordinary fish with an extraordinary prince and an extraordinary doctor of medicine resulted in the publication of what has been called the first geological treatise, and in the elucidation of three basic principles of historical geology. A brief introduction to each of the three characters in this strange cast is thus in order.

Carcharodon carcharias, as the great white shark is known to zoologists, is one of nine man-eaters among the more than 250 kinds of sharks living today. Adult specimens average around 18 feet long, although individuals estimated to be twice this length have been sighted. The behavior of this species in the presence of men or boats is unpredictable: authenticated accounts of the fish's fleeing in apparent fright from divers contrast strangely with equally reliable records of its damage to fishing boats by repeated ramming and of occasional savage and fatal attacks upon swimmers. Though not abundant today, the white shark is wide-ranging throughout the temperate seas of both hemispheres. Judging by the gape of the specimen examined by Steno, the entire animal was 14 or more feet long, and probably weighed about a ton (*1*)*.

Nicolaus Steno

Ferdinand II was 18 years old when he took charge of the government of Tuscany in 1628. The wealth and political influence of the Medici family, already in decline before the death of his father in 1620, continued on the downgrade through the half century of his rule. During the years of his minority, affairs of state were managed, or rather mismanaged, by his mother and grandmother acting as co-regents according to the will of Cosimo II (1590–1620). In defiance of this will, however, the treasury of the house was squandered and the regents adopted a submissive and subservient attitude toward the Holy See. Despite the ecclesiastical influence prevailing in the court, however, Ferdinand's mother, the Grand Duchess Maria Maddalena, sent each of her sons to study physics with Galileo (*2*).

Ferdinand is remembered more for his generous disposition, his

*Numbers in parentheses refer to the Notes in the back of the book, beginning on page 223.

love of peace, and his cultivation of the arts and sciences than for his administrative ability. Among other accomplishments he sponsored the work of the *Accademia del Cimento,* founded in 1657 by his brother Leopold. Over the period of a decade, members of this "Academy of Experiment" developed new and reliable instruments for the measurement of the temperature and humidity of the atmosphere. The Grand Duke himself is credited with the invention of the condensation hygrometer, a device for measuring the humidity of the air depending on the formation of dew on surfaces artificially cooled. He was also the first to overcome the sensitivity of tubular thermometers to changes in barometric pressures by hermetically sealing the upper ends of tubes partly filled with liquids.

This active interest in the advancement of science on the part of Ferdinand made possible Steno's entry into the scientific circles of Florence. In 1667 Steno was given a monthly pension and was assigned rooms in the Palazzo Vecchio (*3*).

Steno was born in Copenhagen in 1638. His father, Sten Pedersen, came from a family of preachers but chose goldsmithery as his trade. Pedersen's work with precious metals and stones must have been good, because his shop was called upon to supply the courts of the Danish kings Christian IV and Frederick III.

After training in a Latin school, Steno was admitted to the University of Copenhagen in 1656. There he studied with Thomas Bartholin (1616–1680), professor of anatomy, whose fame rests mainly on his pioneer studies of the human lymphatic system. After some three years of study Steno went on an educational journey to Holland. Early in his residence at Amsterdam he discovered the excretory duct of the parotid gland to the mouth. Textbooks of anatomy still refer to this vessel as "Steno's duct."

As a student at the University of Amsterdam Steno published and defended a scholarly disputation on thermal springs, their mineral content and sources of heat (*4*). Thus his first publication was on a geological subject, though his primary interest in spring waters probably lay in their real and supposed medicinal values. In any case he continued anatomical and medical studies at the University of Leiden, where he conducted research on glands over a period of two years.

Steno did not tarry at Leiden to receive his M.D. degree, which was awarded *in absentia* while he was conducting studies of muscles and embryos in Paris. There he had the good fortune to become a member of a circle of research-minded savants led by Melchisédech Thévenot (c. 1620–1692). One of the founders of the French Academy of Science, Thévenot was a physician by training but was also a master of Oriental languages, an inventor of surveying instruments, a bibliographer, and after 1684 Librarian of the Royal Library. In 1665 Steno prepared for the Thévenot group a lecture on the anatomy of the brain, a work that has since become famous and is still in print (*5*).

In mid-September of 1665 Steno left Paris and set out for Italy. What led him to undertake this arduous journey of more than 1,500 miles is not known. Some have suggested that he wished to become acquainted with Galileo's group of students in Florence. However that may be, in late March or early April of the following year he arrived at Pisa, the winter residence of the Medici Court. There his attachment to Ferdinand II began.

When Steno received the shark's head, he was preparing for publication a lengthy report on the geometry of muscles. In 1667 his account of the shark appeared as a supplement to this longer work under the title *Canis carchariae dissectum caput* ("The head of a shark dissected") (*6*). Here we are reminded that the significance of a scientific document is not to be gauged by its length, for while Steno's geometry of muscles is remembered mainly as an egregious example of how mathematics can be misapplied to biological problems, the supplement stands as a geological classic. Moreover, that part which is so highly esteemed is not the description of the specimen at hand but a digression into the origin and significance of certain curious objects known from antiquity as glossopetrae or "tonguestones."

Glossopetrae are flattish objects, in side view resembling equilateral triangles, with heights ranging to four inches or more. The thickened bases taper to pointed or narrowly rounded ends along sides with serrate edges. Faces are commonly enameled and shiny, and their colors range from light gray to almost black. Viewed from above, the concave side bears a gruesome resemblance to a tongue that has been turned to stone. Tonguestones are found embedded in rock or lying on the surfaces of rocks that contain them. The overall appearance of these objects is sufficiently striking to attract the attention of anyone interested in natural curiosities.

Though glossopetrae have been found in various parts of the world, those from the Maltese Islands have played the largest role in folklore and in the science of paleontology (*7*). By the early part of the seventeenth century specimens from Malta were widely scattered over Europe in private collections and museums. Steno had probably seen the Maltese tonguestones that were in collections at the University of Copenhagen when he was there as a student. His professor, Bartholin, had visited Malta but could make no final decision as to the origin of these curiosities.

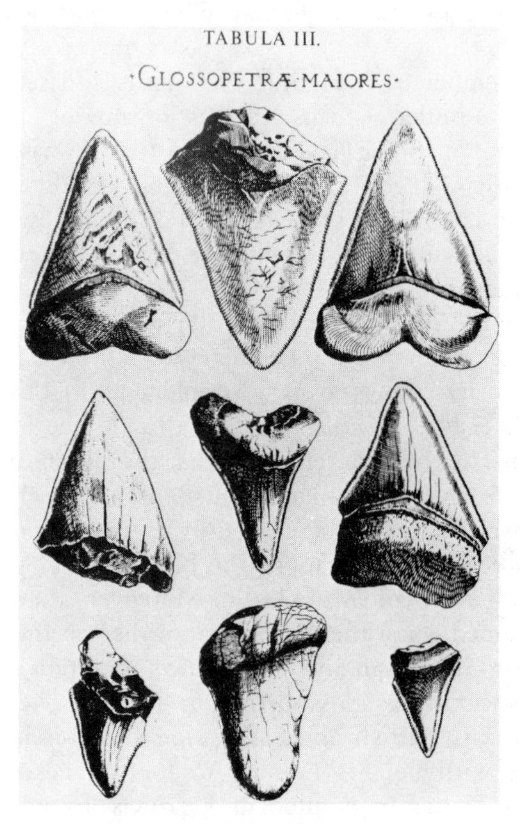

Glossopetrae (*From an imprint of a plate by Mercati*)

An early, perhaps the earliest, reference to tonguestones is found in the last book of Pliny's *Natural History*. In the following passage Pliny seems willing to grant that these stones have fallen from heaven, but questions their power to excite sexual desires or to stay flatulence. The quotation is from Philemon Holland's text of 1601, earliest translation of the *Natural History* into English.

> Glossi petra resembleth a mans tongue, and groweth not upon the ground, but in the eclipse of the moone falleth from heaven, and is

thought by the magicians to be verie necessarie for pandors and those that court faire women: but we have no reason to believe it considering what vaine promises they have made otherwise of it; for they beare us in hand, that it doth appease winds. (*8*)

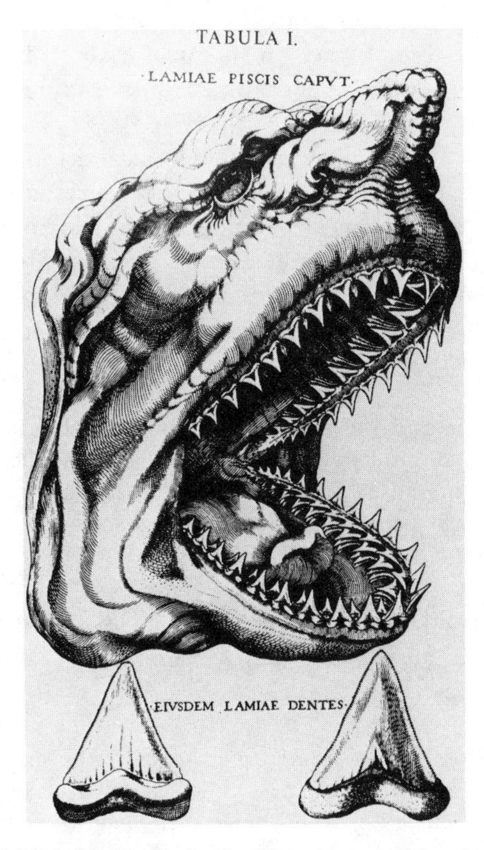

The head and teeth of a modern shark (Plate prepared by the papal physician and naturalist Michele Mercati (1514–1593) and reproduced in Steno's shark's-head treatise)

A distinctively Maltese hypothesis for the origin of tonguestones relates them to an unscheduled visit of Paul the Apostle to the island in A.D. 59. Shipwrecked near the bay that now bears his name, Paul was bitten by a viper. To the amazement of the natives, he suffered no ill effect (*9*). Paul's curse upon all the snakes of the islands deprived them of their venom and turned their teeth into the tonguestones. Another version maintains that replicas of the tongues and not the teeth of the snakes grew in the earth following the miracle worked by the Apostle. Hence the names Maltese tonguestones or St. Paul's tongues. Certain country people of Malta still believe that the "Ilsien San Pawl" has the power to cure human ailments. (*10*)

According to a third tradition, maintained until late in the seventeenth century, tonguestones are simply minerals that have grown in the earth. Their resemblance to tongues or teeth is coincidental or the result of natural or supernatural forces acting in a frolicsome manner.

Lusus naturae ("sport of nature") was a catch-all term commonly applied to fossils during the seventeenth century and before. In October of 1663, James Long, a member of the Royal Society of London and a writer on agriculture, wrote a letter to Henry Oldenburg, Secretary of the Society, telling about finding some objects resembling the shells of snails ("screwstones") and clams embedded in sandstone.

> I digged for some of the screw stones and found that the sandy stone . . . was full and as it were confected of sundry shells of sea fish although it bee on a hill in an inland country. There are cocles and musles and long wrought periwincles amongst thes screws so that it seems nature is gamesome in that place. (*11*)

Also in the Oldenburg correspondence is a letter of 1668 from René François de Sluse, a mathematician of Liège, describing objects resembling shellfish.

> Lately some stones were brought me; I enclose the shape of one of them sketched from either side. . . . . They are dug, I am told, five miles from here in a hill . . . . It is strange that such little creatures should be found, hardened by a petryfying juice, so far from the sea; unless indeed they are the results of nature's playfulness. (Translated from the Latin original by A. R. and M. B. Hall.) (*12*)

When Steno examined the teeth of the shark from the Ligurian Sea he recognized their strong resemblance to tonguestones, although they were only about half as long as the largest glossopetrae from Malta. The supplement describing his dissection of the shark's head is mainly a defense of the proposition that the tonguestones have grown not in the ground but in the heads of sharks. Of course the main problem was to explain how the teeth got from the shark's head into the solid rock that encloses them. Or in more general terms how has one solid body come to be inside another solid body?

Steno begins his argument by stating what is known, from observation and not hearsay, concerning bodies which like the tonguestones resemble parts of aquatic animals. They occur in loosely consolidated as well as hard rock. Commonly the rock is made of superimposed layers. In the case of clay these bodies are usually more abundant on the surface than below. In rocky ground these bodies are attached to the rock as though they had been set in plaster. Whether taken from hard or soft rock, these bodies are not only very much like each other but also closely similar to the animal parts they resemble. Bodies resembling shells of aquatic animals have the same kinds of ridges, laminar internal texture, curvature and windings of cavities, and in the case of bivalves the same hinge structures as are found in their living counterparts. These bodies may be whole or fragmented. Some resemble the broken shells of mussels and scallops, or aggregates of deformed oyster shells cemented together to form hard lumps. Numerous tonguestones of different sizes have been found embedded together in the same matrix. Given these facts, Steno then proceeds cautiously, through a series of six "conjectures," to the conclusion that the bodies in question not only resemble but in fact are parts of organisms.

Steno begins his chain of conjectures with an attack on the then popular idea that the bodies in question grow in the ground. "The ground from which the parts of animals were dug out does not seem to produce these bodies today," Steno notes. Indeed, in the case of soft ground such as clay, these objects appear to be in process of destruction rather than growth. As for their usual concentration on the surface of the ground, this is due to the removal of the matrix by rainwash before the bodies could be completely destroyed in the soil. In hard ground (such as limestone) these bodies are found throughout rock of the same consistency, surrounded on all sides by hard material which would have to be displaced if such objects were growing today.

Steno concedes that certain kinds of solid bodies grow in the ground. Tree roots, for example, wedge their way into cracks and crevices of hard rocks and walls; but they must accommodate themselves to the spaces available in the hard substances they invade, and so become twisted and compressed in ways different from root systems in soft ground. By contrast, each of the bodies in question maintains the same shape whether in soft ground or hard. Hence it would appear that the matrix was not firm when these bodies were formed.

Indeed, he continues, there seems to be no objection to the belief that the matrix containing the bodies resembling parts of animals was originally a sediment gradually accumulated from water at a time when the present land was flooded. Inundations of lands in times past may have been accomplished by either a rise in sea level or a lowering of the ground. Sudden changes in the height of the land above sea level have taken place during violent earthquakes, such as some in Asia Minor described by Tacitus. And, according to Scripture, all things were covered with water, first at the beginning of Creation and later at the time of the Flood.

To say that the matrix originated from the settling of sediment suspended in water poses no great problem, Steno contended. We know from experience that sand and clay particles are easily mixed in agitated waters or in strong winds. The layering seen in the ground containing the bodies in question itself suggests that the matrix has formed by successive settling of sediment from water.

How then did the remains of aquatic animals become entombed in this sediment? Steno suggests that this process may be seen going on today in flooded bottoms of abandoned quarries. Here sediment settles gradually through the stagnant water, covering the parts of dead animals left on the bottom, while the bottom-living animals produce new progeny which in turn will be buried under newer layers of sediment.

The fact that the present mineral composition of the bodies in question may not be the same as found in the parts of living animals which they otherwise resemble presents no great problem. Invisible solids are present even in clear water, either dissolved or suspended as fine powders. These substances circulate through underground passages and may introduce new minerals into the buried remains of animals.

Steno's conclusion that the tonguestones actually are the teeth of sharks is expressed with less confidence than the logic of his arguments might seem to justify. "While I show that my opinion has the semblance of truth," he states, "I do not maintain that holders of contrary views are wrong . . . Nature in her operations achieves the same end in various ways." Was Steno writing this disclaimer, and did he make earlier reference to Scriptural versions of floods, as formal exercises necessary to secure the Church's consent for publication? Or was he hoping to leave open for reconciliation two apparently different models of earth history, one based on the authority of the Holy Scriptures and defended by the Church, the other developing through his own investigations of rocks and fossils? We may never know the answer, but in any case the censors found nothing contrary to the doctrine and moral precepts of Catholicism in this work, which was allowed to be published in Florence in 1667. Whether truly diffident or merely politic, Steno ended his treatise by exploring the implications of his hypothesis.

> If we believe the (historical) accounts, new islands have emerged from the sea; and who knows where Malta's cradle was situated? Perhaps formerly when this place was submerged in the sea it was the haunt of sharks, whose teeth in times past were buried in the muddy sea-bed; then afterwards, when it had changed its level by a sudden ignition of subterranean emanations, these sharks' teeth are found in the middle of the island. (*13*)

The search for Malta's cradle, and the cradle of the Italian mainland, would occupy Steno's attention for many strenuous months ahead.

CHAPTER THREE

Solids within Solids

Where there is matter, there is geometry.
JOHANNES KEPLER

After completing his work on the geometry of muscles, Steno began to broaden the range of his studies. Although he remained interested in anatomical investigations, his concerns turned more toward the characteristics and origins of minerals, rocks, and fossils. In 1667 and 1668, his geological field trips took him to all parts of Tuscany, down through the mountains of central Italy, and westward onto the island of Elba. His collections encompassed a wide variety of natural objects including crystals of quartz and other minerals, ores of mercury and silver, petrified wood, specimens of volcanic and sedimentary rock, and fossilized mussels and snails. This collecting was not done merely to assemble an array of natural curiosities. We know from Steno's correspondence that he was planning a major work on what we should today call the principles of historical geology, with special reference to Tuscany and nearby regions. Two major crises in Steno's life, one quickly following upon the other, would prevent the completion of this major work (*1*).

On November 3, 1667, Steno converted to Catholicism. And on the following eighth of December, the day of his confirmation, a letter arrived from the Danish Court advising him to return to Copenhagen as royal anatomist, by authority of Frederick III.

Steno did not leave Florence immediately. Presumably he wrote to his king, either asking whether the appointment could be made to a Lutheran turned Catholic or petitioning personal religious liberty. In

June of 1668 he must have received some word of reassurance, for on the following July 4 he gave up his rooms in the Palazzo Vecchio. Before setting out for home, however, he struck off a long summary of the treatise he had planned to write. The manuscript, completed in less than two months, was published the following April under the cumbersome title *The prodromus of Nicolaus Steno's dissertation concerning a solid body enclosed by process of nature within a solid* (*2*). A prodromus is a preliminary discourse; evidently Steno still hoped to complete the larger work.

The *Prodromus* consists of an introduction followed by four sections of argument. In the introduction Steno explains to the Grand Duke, his patron, that this work represents only partial payment of a personal debt of gratitude. He regrets that leaving Florence will delay writing the major text, though the journey ahead not only will give opportunities for further investigations, but also will allow him to improve his command of "the Tuscan tongue" in which he hopes to cast the expanded text. Meanwhile his summary in Latin must not be taken as a complete account of his observations and conclusions. "Sometimes," he states, "I shall report . . . conclusions, and again observations, as may seem best, in order to explain the chief points briefly . . ." (*3*).

Steno then proceeds to develop an historical and philosophical background for his investigations. Efforts to solve problems concerning the natural world, he observes, have generated more questions than answers, for two reasons. Either the questions for which answers are sought do not get to the heart of the problem at hand, or else the investigators have not distinguished what is solvable from what is not. Thus with regard to bodies resembling parts of marine organisms but enclosed in rocks of the land, the Greeks framed the question: "In what way have marine bodies been left so far from the sea?" A prior and more fundamental question that was never posed asks whether these bodies are produced in any place *other* than the sea. Furthermore, to understand the origin of these objects we must determine whether they originated in the places where now they are found; and to know this we must first understand the *manner* in which these bodies are formed; and, finally, to know that we must have some comprehension of the nature of matter itself.

As for the nature of matter, Steno asserts that several fundamental questions must be regarded as insoluble in the light of contemporary knowledge. He will offer no judgment as to whether the ultimate particles that make up a natural body undergo any sort of change as the body itself changes in shape. He will not speculate on the presence or absence of minute voids between these particles, nor on the characteristics of these particles beyond the evident facts that they occupy space within and may lend firmness to the bodies they form. He will, however, accept the propositions that natural bodies are made of imperceptible particles, that in fluids these particles are in constant motion, and that in solids the particles may sometimes be in motion, especially when the solids are being formed. Through the aggregate of imperceptible bodies that form natural bodies may pass forces generated by the magnet, fire, and in some instances light. As for the origin of the motion of the particles, nothing is known, though the first cause of motion must have been divine. Changes in the destination of moving particles to form new bodies may be due to natural, artificial, or divine causes.

Steno then proceeds to set down three principles to guide research on natural solids. These principles provide answers to three basic questions. Given a solid body enclosed on all sides by another solid body, which of the two first became hard? How is the manner of origin of some specific solid body to be determined? Taken together as a class of objects, how are natural solids formed?

His answer to the first question is that of the two bodies under examination the one that first became hard is the one that has left the impression of its surface on the other. For example, a scallop shell will leave the imprint of its external ridges and furrows on the mud in which it is buried. And if the mud hardens to mudstone the ridges and furrows earlier imprinted on it will tell that the shell was hard while the substance of the rock was still plastic.

To the second question Steno offers the proposition that if two solid bodies resemble each other with regard both to external form and to internal structure they very likely originated in the same way. This was the basis for Steno's reasoning about the origin of tonguestones.

As for the origin of natural solids as a class, Steno asserts that all have been produced from fluids—layers of rock, nodules and concretions within rock layers, minerals filling veins, crystals within rocks, the skeletons of organisms embedded in rock, and even the tissues of plants and animals and the callus uniting broken bones—all are produced from fluids of some sort, whether of air, water, or the body fluids of organisms.

Steno then moves from generalities to particulars. That the

layered rocks of the kind so commonly found in Tuscany were formed by sediments settling out of water is indicated by their texture and composition. The larger particles of sediment are generally found toward the bottoms of these layers, an expression of the laws of gravity which dictate that in a mixture of sediment and water the larger and heavier bodies will fall to the bottom before the smaller and lighter. The nature of the bodies enclosed in the strata tells whether the water in the original suspension was fresh or salt. If the strata contain deposits of salt, skeletons of marine organisms, ships' timbers, or sediments like those now found on sea floors, we may infer that they originated at the bottom of the sea. On the other hand, if the strata contain trunks and branches of trees or parts of other terrestrial flora, presumably they accumulated in fresh water.

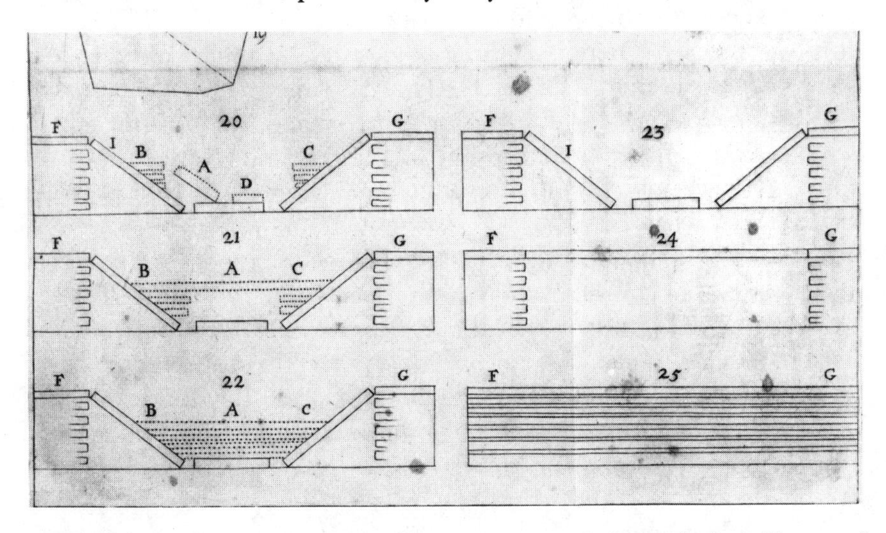

Steno's cross-sections illustrating his conception of the geological history of Tuscany. For explanation, see text. (From an original copy of the Prodromus *in the DeGolyer Western Collection, Southern Methodist University)*

The central idea in Steno's discussion of stratified rocks is that stacks of strata have accumulated *layer by layer* and not all at once. One layer of sediment forms on the bottom of a sea or lake, and at the end of this event there is nothing but water above that layer. New quantities of sediment are introduced into the water and there settle to form a second layer on top of the first, and so on until the sequence of strata is completed. Steno's generalization on this matter is that "at the time when any given stratum was being formed all the matter resting upon it was fluid, and, therefore at the time when the lowest stratum was being formed, none of the upper strata existed" (4). This is an early formulation of *the principle of superposition of strata*, upon which much of what we think we know concerning the history of the earth and the moon is based. A modern phrasing of this principle might be: "In a sequence of sedimentary strata, as originally deposited, any stratum is younger than the stratum upon which it rests and older than the stratum that rests upon it."

Simple? Yes. We know from everyday life that in any kind of stack or pile, as originally stacked or piled, the objects at the bottom got there before those at the top. In a bridge hand freshly dealt, the first card that left the dealer's hand will be at the bottom and the last at the top of the stack. We see a rug on a floor and reasonably infer that the floor was there before the rug was laid.

Obvious? No. There is nothing obvious about inferring time from solid geometry, and that is precisely what Steno did. "At the *time*," he says, "when the lowest stratum was formed, none of the upper strata *existed*." By this conception, a stack of solid strata translates into a sequence of intangible events, to become as it were pages in the history of nature.

Another footnote to the principle of superposition. Observe that when Steno said in the above context that "none of the upper strata existed," he did not deny that the *particles* in the upper strata were in existence somewhere. He means only that the particles in these strata had not gathered together as a class when the lowest stratum was formed. Superposition tells nothing about the relative ages of the particles in strata nor about the absolute ages in years of the strata themselves; the principle only yields "older than" and "younger than" relationships among beds disposed in stacks.

Steno's conception of strata as sediments of fluids led him to two other generalizations which have been useful in inferring events from rocks. The bottom of the first stratum in a stack of strata should reflect the irregularities of the solid bottom on which it came to rest, but the upper surface would be more even. With continued deposition the ups and downs of the bottom would cease to be reflected in the attitude of newer strata, which would lie flat and essentially horizontal. This is the *principle of initial horizontality of strata*. Its application to historical geology was explicitly stated by Steno in these terms: "Strata either perpendicular to the horizon or inclined toward it were at one time parallel to the horizon" (5). Thus when we see exposures of steeply dipping strata we may infer that at some time after the beds were formed they were rotated or tilted out of their original position during one or more events of dislocation.

Steno conceived of a stratum as being a *continuous* solid layer at the time of its formation. The sediment forming the stratum would disperse "over the whole surface of the earth" unless older solid bodies stand in the way. Thus when we see the bared sides of strata we may infer that something has been removed since the layers formed—either the obstruction that prevented the further lateral spreading, or some parts of the strata themselves. This is called *the principle of stratal continuity* (*6*). A trivial application of this principle may be made at nearly any road-cut through a sequence of stratified rocks, where oftentimes we can match the "bared sides" of individual beds that were partly destroyed in constructing the road. An event of road-building would be inferred to have taken place after the strata formed. By the same principle we infer events of stream erosion from matching beds on opposite sides of the Grand Canyon.

Admittedly Steno was oversimplifying and overstating when he proposed that all strata in a sequence, except the oldest, were originally horizontal, and that a single surge of sediment could spread over the whole globe. Not even the greatest known volcanic eruptions have covered the earth with ash, and under certain conditions sediments may form with initial inclinations of more than 20° to the horizontal. But with reasonable qualifications the principle stands.

Steno recognized that the contours of the landscape undergo continuous change. Strata may be thrust up by explosions of flammable subterranean gases or by expulsion of subterranean air attending the collapse of caverns. The collapse of caverns may also result in downward tilting of strata toward the depression produced thereby. Mountains are not all of the same age or origin: some are produced by volcanic eruptions, while others stand as residuals above valleys carved by streams. The peaks of mountains may be raised or lowered, so that it is not necessary to assume that all present mountains existed from the beginning of things.

The long section on minerals in the *Prodromus* we can pass over without summary for present purposes, but not without mentioning that Steno made an important observation concerning the angles between adjacent faces on the six-sided prisms exhibited by quartz crystals. Regardless of the sizes of different crystals or the relative sizes of the prism faces on the same crystal, the faces always come together at the same angle. This is a singular example of a broader principle which states that like faces on all crystals show a constant angular relationship. This *principle of the constancy of interfacial angles* has sometimes been called Steno's Law (*7*).

Returning to solids resembling parts of animals, Steno recognizes three different classes. First, there are those objects so well preserving all the characters of their modern counterparts that their organic origin can scarcely be questioned. In this group fall certain fossil shells retaining parts of their original colors, and oyster shells perforated by borings and closely resembling recent specimens found along the Italian shores. In a second class of these objects belong fossils which have been modified from their original condition by changes in color and weight as a result of petrification. Finally there are those objects which preserve only their original shapes, and have been replaced in substance by new mineral matter.

Steno concludes his *Prodromus* with an account of the geological history of Tuscany based upon the principles set forth in the preceding sections (*8*). To illustrate the sequence of main events he drew six cross-sections to show stages in the development of the present landscape. He begins with the present and works backward in time. His Fig. 20 shows the present situation in Tuscany: a sequence of softer strata (layers bordered by dotted lines) mostly horizontal and either resting upon or abutting against a sequence of harder rocks (solid lines) which form the present mountains. By the principle of superposition, the softer strata must be younger than the harder. By the principles of initial horizontality and stratal continuity, the softer strata must originally have formed a solid stack filling lowland areas between uplands formed of the harder strata, as shown in his Fig. 22. By application of these same principles, one must envision an original situation, prior to the deposition of the softer strata, when the harder strata formed a solid, horizontal stack (his Fig. 25).

The problem now is to account for the tilting of the strata. Perhaps drawing upon his experiences in exploring grottos and caverns, Steno guessed that after the older strata formed they were partly destroyed by the corrosive action of underground water or fire, leaving the uppermost layers standing as natural bridges over subterranean voids (his Fig. 24). Eventual collapse caused the downtilting of strata toward the center of the cavern. By the same token a subterranean cavity formed by dissolution of part of the younger and softer strata produced the cavern shown in his Fig. 21; and the collapse of the roof produced the general contours of the present landscape.

Thus, according to Steno's interpretation, Tuscany has twice been flooded by the sea, has twice been dominated by level plains following retreat of the sea, and twice has been dislocated by collapse of subterranean hollows.

Steno realized that Christians might be alarmed by the novelty of his views. Accordingly he equated the second time of flooding and deposition of the younger sediments to the Universal Flood of the Scriptures. The older set of marine strata he attributed to deposition in the universal ocean after the second day of Creation, according to the Mosaic account. This ordering left but little time between the two events of flooding. For, by Steno's calculations, the old Etruscan city of Volterra must already have been about 3,000 years old in A.D. 1669. Volterra is built on a mountain made of the younger sequence of sediments. Hence the Flood must have subsided no later than about 1331 B.C. But Lydia was in existence about a thousand years before Volterra; hence 2331 B.C. is a more probable date for the last emergence of the lands. This would leave only about 1,650 years for the interval between the earlier and later floodings, allowing for a date around 4000 B.C. as the date of Creation. Having correlated with Biblical events the two episodes of flooding in Tuscany, Steno resolutely tried to do the same for the other four stages of development.

This exercise at reconciling geological history with Christian dogma succeeded in gaining the Church's imprimatur. The Vicar General of Florence directed that Vincenzio Viviani examine the manuscript and report whether anything in it might be construed as contrary to Catholic Faith or good morals (*9*). The mathematician Viviani, last student of Galileo and leading academician of Florence, promptly replied that the work was essentially a dissertation on the whole of physics and that its author demonstrated an unreserved respect for the Faith. Viviani's prestige was undoubtedly a factor in gaining the Church's consent for publication, finally given after a delay of about three months. Even so, considering the obvious incompatibility between the generous requirements of time implied by Steno's stratigraphic principles and the strictures on time imposed by literal interpretation of Scripture, some have wondered how Steno escaped censure.

The *Prodromus* completed, Steno set out for Denmark late in October of 1668. No one could have guessed the destination from the itinerary. From Florence he headed southward to Rome and Naples, then across the Apennines and along the Adriatic to Venice and then to Bologna, all along the way making geological observations and conferring with other scholars. After a stay of three months in northern Italy he travelled for half a year in Switzerland, Austria, and Hungary. From Vienna he journeyed through Bohemia and then moving westward across Germany he arrived in Leiden late in 1669. There he learned that Frederick III had died, so that it no longer seemed necessary for him to go on to Copenhagen.

In May of 1670 Steno was informed that Ferdinand II was ill. Accordingly he hastened back to Florence, arriving there the following month only to find that his patron had died. The new Grand Duke, Cosimo III, provided his old tutor with quarters and urged him to continue his scientific studies. But Steno's interests had turned away from science. While in Holland he had begun a theological disputation with the pastor of a German Reformed Church in Amsterdam. During the first half year in Florence he continued to develop his theological arguments. Disinterest in science soon turned to disaffection, judging by a letter written to a friend in 1671 in which he declared that we actually know nothing about the origins or original structures of natural objects.

Early in 1672, Steno's residence in Florence was again interrupted by an order from the Danish King, this time Christian V. Steno was to return to Copenhagen as soon as possible to assume duties as royal anatomist with a generous pension. Steno promptly complied, arriving at Copenhagen in July. There he delivered lectures to students and performed dissections on animal and human bodies. In earlier years he had hoped for a professorship at his university, but fulfillment of the wish did not match his expectations. As a Catholic he could not enjoy all the privileges usually accorded a professor. Nor, would it seem, could he refrain from introducing metaphysical references into his lectures. From the preface to one of his public lectures comes a line that has been widely quoted: *Beautiful is that which we see, more beautiful that which we know, but by far the most beautiful that which we do not comprehend* (*10*). This is taken to mean that natural science, the science of mind, and Christian Faith are a triad on an ascending scale of significance. Steno had at last set his values in order. After two years as anatomist he resigned his position and returned to Florence as tutor to Crown Prince Ferdinand.

Steno spent the remainder of his life in ministerial work. He was ordained a Catholic priest in April of 1675. He took the vow of voluntary poverty and adopted as his motto *De claritate in claritatem*—from light to light. His ascent in the officialdom of the church was rapid. Scarcely more than two years after his ordination, Pope Innocent XI named him Vicar Apostolic for Hannover, Germany, and less than a month later called him to Rome for consecration as Bishop.

Stationed at the small ducal court of Hannover, Steno presided over a vast but poor diocese which included northern Germany and Scandinavia. He lived in self-imposed poverty, giving to the poor all his belongings including his silver crucifix and bishop's ring. After Hannover fell to a Protestant duke, Steno moved to Münster as Auxiliary Bishop. But when the new bishop bought his way into office Steno showed his contempt for traffic in things sacred by refusing to attend the mass of celebration. Leaving Münster in 1683 he went to Hamburg as Vicar Apostolic. There his frugality offended even his parishioners, some of whom actually threatened to do him bodily harm. In 1685 Steno undertook a missionary assignment to the ducal court at Schwerin. There, after what has been described as an ordeal of poverty and fasting, he died on November 26, 1686. Cosimo III ordered the body to be returned to Florence and buried in the chapel of San Lorenzo.

The *Prodromus* had an immediate impact on British scientists, whose successors in the centuries to follow did so much to advance knowledge of the earth's history. Less than two years after this work was published at Florence, Henry Oldenburg not only issued an English translation but also advertised the book in the *Philosophical Transactions* along with a four-page summary of its contents (*11*). That Oldenburg placed a high value on Steno's ideas is sufficiently indicated by the hours he must have spent in "Englishing" the long Latin text. He also knew that the importance of the book was obscured by its title, and so he entered the following subtitle of his own design: "Laying a Foundation for the Rendering a Rational Accompt both of the *Frame* and the several Changes of the Masse of the EARTH, as also of the various Productions in the same."

The *Prodromus* was read and commented on by a number of Steno's learned contemporaries, including Martin Lister, John Ray, Robert Plot, and John Woodward (*12*). Not all agreed with Steno's ideas; Lister, for one, questioned that petrified shells were truly the remains of organisms. But no one, to my knowledge, questioned the propriety of Oldenburg's subtitle: a foundation for reconstructing the history of the earth by the exercise of reason had indeed been laid.

For his contributions to anatomy and geology, Steno has received his due share of posthumous honors, not the least of which are symbolized by recent translations of all his scientific writings into modern languages, with critical commentaries appended in most instances. What Steno might have regarded as the highest honor imaginable began to materialize in 1938. In that year, on the three hundredth anniversary of his birth, a group of Danish pilgrims obtained the consent of Pope Pius XI to gather documentation on Steno as a first step in beatification. Steno's voluminous theological writings were brought together and published. The process of canonization began in 1953. Steno's coffin was raised from the crypt and opened. His remains were carried in solemn procession through the streets of Florence and then were placed in an ancient sarcophagus at a chapel in the transept of San Lorenzo (*13*). When I visited there in 1964, the sarcophagus was covered with handwritten notes and letters. One note, lying open, was prematurely addressed to "Dear Saint Nicholaus," asking that a baby soon to be born might be sound in mind and body.

I wondered then as I wonder now what led this restless and thoughtful man to turn so completely away from science and toward theology. Perhaps this change grew out of frustration at trying to live simultaneously in two quite different worlds: the world of Scripture with its brief and miraculous history, and the world of Nature whose monuments imply indefinite but vast extension of the earth into time past.

Earlier in the seventeenth century, Francis Bacon, who has been characterized as a revolutionary pagan, predicted that any effort to mix science with theology would come to no good end. Writing nine years after the publication of the King James version, Bacon declared:

> . . . some of the moderns have indulged this folly, with such consummate inconsiderateness, that they have endeavoured to build a system of natural philosophy on the first chapter of Genesis, on the Book of Job, and other parts of Scripture. . . . And this folly is the more to be prevented and restrained, because not only fantastical philosophy but heretical religion spring from the absurd mixture of matters divine and human. (*14*)

This brings to mind Thomas S. Kuhn's recent analysis of the structure of scientific revolutions. According to Kuhn, research in normal science is "a strenuous and devoted effort to force nature into conceptual boxes supplied by professional education" (*15*). These conceptual boxes he calls paradigms—models, or examples. "Discovery commences with the awareness of anomaly, i.e. with the recognition that nature has somehow violated the . . . expectations that govern normal science" (*16*). This leads to crises in scientific thought and to the transformations of models which are called scientific revolutions.

Kuhn goes on to say that "Though history is unlikely to record their names, some men have undoubtedly been driven to desert science because of their inability to tolerate crisis. . . . rejection of science in favor of another occupation is, I think, the only sort of paradigm rejection to which counterinstances by themselves can lead. . . . To reject one paradigm without simultaneously substituting another is to reject science itself" (*17*).

Did Steno desert science because he could not tolerate the kind of crisis Kuhn describes? If so, his crisis must have been the more acute because he was not merely engaged in stuffing nature into conceptual boxes built by someone else; rather he was himself constructing the boxes, the paradigms, and the models for a fresh view of nature which implied that the dimension of time past would have to be greatly extended. The question cannot be answered, for the evidence is only circumstantial. Whatever the answer, I think it a credit to Steno's honesty that he neither created a fantastic science nor lived out an heretical religion.

CHAPTER FOUR

Mr. Hook(e)

> The struggle for knowledge hath a pleasure in it like that of wrestling with a fine woman.
>
> GEORGE SAVILE (LORD HALIFAX)

In the *Philosophical Transactions* of the Royal Society of London for February 10, 1667/8, Henry Oldenburg reviewed Steno's treatise on the geometry of muscles. Referring to the part on the dissection of the shark's head, he noted that in the controversy concerning the origin of the Maltese tonguestones, Steno "takes their part, who maintain that those and divers other substances, found in the earth, are parts of the Bodies of Animals, and endeavours to prove, that such sorts of Earth may be the sediments of water, and such Bodies, the parts of Animals carried down together with those sediments, and in progress of time reduced to a stony hardness." To this summary Oldenburg felt obliged to add a footnote, as follows.

> This Subject *Mr.* Hook *hath also discoursed of at large in several of his publick Lectures, founded by* Sir John Cutler; *which Lectures he read about Two years since in* Gresham College, *in the presence of many Learned and Curious persons; which also had been long since made publick, had not other indispensable affairs hindred him from taking care of the Press: where he hath not only shewn the origin of these* Glossopetrae, *but of all other curiously figur'd Stones and Minerals, together with that of Mountains, Lakes, Islands, etc. though from a somewhat differing* Hypothesis, *of which the curious may shortly receive a further Account. (1)*

The Mr. Hook of this footnote is of course Robert Hooke, who continues to be acknowledged in footnotes wherever such diverse topics as optics, astronomy, microscopy, atomic theory, horology, thermodynamics, paleontology, botany, oceanography, geology, or navigation are detailed in an historical context. Perhaps he is most widely known for his discovery of the law of elasticity—Hooke's Law, the proposition that deformation of elastic bodies is proportional to the force applied. Less well known are his formulations of atomic and kinetic theory. Hooke held that the properties of matter are to be understood in terms of the motions and collisions of atoms, and that what we call heat is no more than the property of a body arising from the motion or agitation of its parts. His experiments in optics contributed to the wave theory of light and further elucidated the phenomenon of diffraction.

Hooke was not, however, concerned wholly or even mainly with theory. He was a practical man, an inventor, a "mechanick genius" in the estimation of his contemporaries. He has been credited with the invention of the universal joint, the wind gauge, the wheel barometer, a water sampler for collecting sea water at measured depths, the anchor escapement of pendulum clocks, and the balance spring for watches. Working with Robert Boyle, he built a device for evacuating air from a glass chamber, and the experiments conducted with the aid of this "pneumatic engine" demonstrated that air is necessary for combustion, for the respiration of animals, and for the transmission of sound. Through microscopes, improved according to his own design, Hooke examined a variety of artifacts and natural objects, discovering among other matters that the most pointed needle does not actually end in a point and that mites, lice, and fleas possess a peculiar beauty of their own. Examining thin slices of cork he discovered the cellular structure of plants, and named the tiny compartments "cells" because they reminded him of the clusters of small rooms occupied by monks.

These accomplishments, and the fame that came with them, could hardly have been predicted from consideration of his childhood and early upbringing (*2*). Hooke was born on July 18, 1635, at the town of Freshwater on the Isle of Wight, where his father was curate of the Church of All Saints. The Reverend John Hooke, hoping that his son would also become a clergyman, served as tutor in subjects he considered appropriate to the ministry. But the son was more interested in drawing, and in making mechanical toys; by the time he was thirteen he had built from wooden parts a clock that would run, and had constructed a model ship that would fire off small guns as it sailed near shore. In 1648 the father died, and young Hooke was sent off to London with an inheritance of £100 to study art as apprentice to Peter Lely, the portraitist. But the smell of paint made the boy's head ache, and so he left with his hundred pounds and enrolled at Westminster School, where he came under the influence of the Headmaster, Dr. Richard Busby, who was famous for his skill as a teacher. Recognizing Hooke's unusual mental endowments, Busby coached him in studies of geometry, in other branches of mathematics, and finally in mechanics, the subject for which Hooke ever afterward showed an especial fondness. Not all of his training at Westminster was mathematical and scientific; following the classical tradition of learning, he studied Latin and Greek and acquired some competence in Hebrew. Also he studied music and learned to play the organ. Though Hooke possessed a natural vigor, his health was always delicate, and when he was about sixteen years old a growing deformity manifested itself in a crooked posture which marked him for life.

At the age of eighteen Hooke obtained a position as chorister at Christ Church, Oxford. Dr. John Wilkins, then Warden of Wadham College, soon became his mentor and encouraged the young man in further studies of mathematics, astronomy, and chemistry. After Robert Boyle set up his chemical laboratory at Oxford in 1654, Hooke became his assistant in experiments. Soon after Wilkins and Boyle had organized a club to discuss and conduct scientific experiments Hooke became a member of the group and a participant in their weekly meetings. All members of the club were advocates of the new experimental philosophy which sparked the scientific revolution of the seventeenth century, and most of them were young. Wilkins, the senior member, was 41 when Hooke joined the group at age 20; four others of the leading lights were in their thirties; and five others, including Boyle and Christopher Wren, were in their twenties.

After the death of Cromwell in 1658, most members of the Oxford group moved to Gresham College, London, where they joined with other scholars and scientists in laying plans for a permanent organization of investigators. The eventual outcome of these efforts was the establishment of the Royal Society of London for the Improvement of Natural Knowledge, by charter of Charles II issued on July 15, 1662 (*3*). Later that year Hooke was appointed Curator of Experiments. In 1663 he was elected a Fellow of the Society, and two years afterward was named Curator for life and Professor of Geometry at Gresham College, where he was provided with living quarters.

As though his professorial duties and responsibilities for preparing weekly lectures or experiments for the Fellows were not enough to occupy his time, Hooke assumed the additional responsibility of delivering 16 lectures each year on subjects to be selected by the fellowship. (For these lectures Sir John Cutler had indicated he would pay Hooke £50 per annum.) Then, following the Great Fire of September 2, 1666, Hooke was appointed a city surveyor to determine the boundaries and courses of streets, draw up plans for new buildings, and in certain instances actually to superintend the rebuilding. This architectural effort provided Hooke with additional revenue, and many pleasant opportunities to work with his good friend Sir Christopher Wren, but consumed much of his time over a period of almost a decade.

Between June of 1667 and September of 1668 Hooke delivered lectures on geological subjects before the "many learned and curious" persons attending meetings of the Royal Society. These were the lectures to which Oldenburg referred in his footnote. They were edited by Richard Waller and published in 1705, two years after Hooke's death, along with his writings on other subjects. The title of the geological section of these posthumous works gives a fair indication of some problems which were on Hooke's mind when he was a man in his early thirties. It is: *Lectures and discourses of earthquakes and subterraneous eruptions, explicating the causes of the rugged and uneven face of the Earth, and what reasons may be given for the frequent finding of shells and other sea and land petrified substances, scattered over the whole terrestrial superficies* (4).

In these lectures Hooke began, not with an account of earthquakes and volcanos, as the title might suggest, but with solid geometry. First he identifies two different kinds of natural stony bodies found lying about on the surface of the earth or embedded in the soil and rocks below: those with and those without definite form or shape. Bodies possessing distinctive shapes he called "figured stones," which he again divided into two classes. The first includes crystals of various minerals—rock salt, feldspar, diamond, ruby, etc.—whose elemental figures display "the ABC of Nature's working" and are peculiar to the substances that form them. Bodies of the second class, which he called *petrifactions*, derive their distinctive shapes not from the nature of the stony substances forming them but from an "external and accidental mould." (Here Hooke is using the word *accidental* in the older sense of something not innate but relating to extrinsic causes or forces.) The stage thus set, Hooke would argue that animals and plants have been the extrinsic causes of petrifactions.

As a preface to this argument, Hooke offered an "enumeration of the phenomena," a listing of what can be accepted as already known about petrifactions. He notes that bodies resembling shellfish, both in substance and shape, have been reported from Eurasia, Africa, and other parts of the world. The same holds for bodies resembling the parts of plants. Equally as widespread are bodies made of stone, clay, or earth but in their shapes closely resembling the whole or parts of plants and animals. Bodies resembling parts of marine organisms are found in rocks forming the highest mountains in the world, in places now hundreds of miles distant from any sea. Petrifactions abound in certain hard layers of limestone, marble, or flint, both where these layers lie upon or very near the surface of the earth and where they have been penetrated in the deepest mines or wells.

Hooke developed his argument as a series of propositions, followed in turn by a matching set of proofs or confirmations. The first group of propositions is directed to the classification of petrifactions and the ways by which they are formed. Most of the bodies in question, he asserts, fall into one or another of three groups: objects in which the original animal or vegetable matter has been turned to stone by filling of the pores with new substance introduced by a liquid; impressions or molds left by the original organic matter on the surrounding materials ("such as heated wax affords to the seal"); and casts of materials introduced into these natural molds. Considering the fact that dead animals and plants usually turn to dust, petrifactions of whatever variety must have required unusually favorable conditions for their preservation.

Hooke was willing to believe that petrifactions have formed under many different natural situations, but he emphasized the capacity of water, fresh or salt, to form stony substances. Familiar examples are the stalactites that hang where water drips from the roofs of limestone caverns, and the durable skeletons which stony corals build from invisible matter dissolved in the sea. To illustrate the ability of sea water to form rock from an aggregate of loose materials, Hooke cited an example on the Isle of Wight. Along the rocky shores of that island masses of sand, clay, and shells sometimes slump from the cliffs and tumble down to the shore, where they are "by petrifying power of the salt water converted into perfect hard compacted stones."

Granted, then, that petrifactions resembling shellfish are indeed of organic origin, Hooke concludes that ". . . a great part of the Surface of the Earth hath been since the Creation transformed and made of another Nature; namely, many parts which have been Sea are now Land, and divers other Parts are now Sea which were once a firm Land . . ." (5). Evidences for former submergence of the lands are widespread in the British Isles, most or all parts of which according to Hooke have had fishes swimming over them. He attributed these dislocations in level to subterranean forces manifest in earthquakes and volcanic eruptions, which have had the additional effect of turning mountains into plains in some places and plains into mountains in others.

In support of these propositions Hooke cited numerous references to changes in level and changes in landscape that had taken place during the sixteenth century as a result of quakes, volcanic eruptions, or the two working together. For example, he mentions the formation of a new volcanic mountain that arose from eruptions in Lake Lucrinus, near Naples, during a time of destructive seismic activity in September of 1538. Similar episodes are cited for the historic records of such widely scattered localities as Iceland, Chile, the Eolian Islands, Java, Nicaragua, and Peru. In Hooke's view, the fact that the continents and the floor of the sea are so irregular testifies to the persistence of quakes and eruptions in dislocating and remodelling the surface of the solid earth. But for the activity of these agencies the land would have been leveled by the work of streams and winds which unceasingly wash or blow sediment from hills and mountains into valleys and plains.

Hooke's "proofs" for the organic origin of petrifactions are cast mainly as arguments against the then more popular view that these bodies are sports of nature grown inside the rocks through some "plastic virtue" inherent in the solid earth. Referring to petrifactions that have the same shape and are made of the same substances as found in living shellfish, Hooke concedes that these are indeed works of a plastic virtue, but of one far nobler than most people conceive.

> . . . at the tops of some of the highest hills, and in the bottom of some of the deepest Mines, in the midst of Mountains and Quarries of Stone, etc.

> divers Bodies have been and daily are found, that if we thoroughly examine we shall find to be the real shells of Fishes, which . . . we conclude to have been at first generated by the Plastick faculty of the Soul or Life-principle of some animal, and not from the imaginary influence of the Stars, or from any Plastick faculty inherent in the Earth itself. . . . (*6*)

Hooke then appeals to the metaphysical proposition that nature, as the servant of the Creator, does nothing in vain and nothing imperfectly. Granting this, he asks how then can we believe that petrifactions originally served no useful purpose, and are no more than artifacts of a sportive Nature idly mocking herself? How explain the fact that most petrifactions are broken and imperfect? If these objects "be the apish Tricks of Nature, why does it not imitate several other of its own works?" Why do we not dig out of mines "everlasting Vegetables, as Grass for instance, or Roses?" If petrifactions are produced by rocks, it might be expected that different kinds would grow in different kinds of rock, when in fact the same kind of petrifaction can be found in different kinds of rock. If we should turn up coins and urns in the course of some excavation, would we say that these had been produced by some plastic faculty of the earth? If not, Hooke asks, why do we question the organic origin of sea shells dug out of the ground?

Hooke recognized that many "fossile-shells" bear only a general resemblance to the hard parts of modern species, and that some are quite unlike anything known to be living today. Detractors of the organic theory made a strong point in emphasizing the differences between certain petrifactions and their alleged modern counterparts. Hooke countered objections of this kind by insisting that the resemblances between the questionable fossils and living species outweigh the differences. Taking as an example the large "serpent-stones" (ammonites) found around Portland, he pointed out that their manner of coiling and their division into chambers by transverse partitions are essential characteristics of the living chambered *Nautilus*, so that the animals that produced these fossils must have been *Nautilus*-like regardless of incidental differences in the geometry of their shells. This proposition immediately raises the question of why no ammonites are found in the oceans of today, and to this Hooke offered two possible answers: they may still be living in some remote and unexplored part of the world, or more likely they may have died out at some time in the past.

Hooke's illustrations of ammonites (Reproduction of Plate 1 in Lectures and Discourses of Earthquakes, *from an original copy in the DeGolyer Geological Library, Southern Methodist University)*

Hooke went on to suggest that species are neither immutable nor immortal—a proposition that must have shocked his audience. He proposed that "there have been many . . . Species of Creatures in former ages, of which we can find none at present; and that 'tis not unlikely also but that there may be divers new kinds now, which have not been from the beginning" (*7*). Extinction or modification of species would most likely occur, he submitted, during times when whole countries sank below the sea or were raised anew from the depths.

Thus in Hooke's view petrifactions are not mere natural curiosities; they are records of the natural history of the world.

> Now these Shells and other Bodies are the Medals, Urnes, or Monuments of Nature, whose Relievoes, Impressions, Characters, Forms, Substances, *etc.* are much more plain and discoverable to any unbiased Person, and therefore he has no reason to scruple his assent: nor to desist from making his Observations to correct his natural Chronology, and to conjecture how, and when, and upon what occasion they came to be placed in those Repositories. These are the greatest and most lasting Monuments of Antiquity, which, in all probability, will far antedate all the most ancient Monuments of the world, even the very Pyramids, Obelisks, Mummys, Hiero-glyphicks, and Coins, and will afford more information in Natural History than those others put together will in Civil. Nor will there be wanting "*Media*" or "*Criteria*" of Chronology, which may give us some account of the time when, as I shall afterwards mention. (*8*)

This passage was read near the end of Hooke's first series of lectures on geology. Unfortunately for us, the promise given in the cryptic last sentence was never fulfilled, and we are left wondering whether or not Hooke had grasped the idea of an orderly succession of different kinds of fossils at different levels throughout thick sequences of strata (*9*).

Hooke's arguments for the organic origin of petrifactions were not as persuasive as he had hoped. After a lapse of almost twenty years he presented a second series of lectures on geology in which he reemphasized and elaborated the main points he had earlier made. In addition he analyzed the mythology of Ovid's *Metamorphoses* and Plato's account of the lost continent of Atlantis, working on the assumption that myths and wonder stories about changes that have taken place on the earth are rooted in fact (*10*). Compared with the earlier lectures, these later ones contain more reflections on theological issues and more references to Biblical passages, especially those relating to the aging earth and to Noah's Flood.

Hooke's conception of the geologic changes that took place in the past would seem to imply that the earth is very old. He believed that at the time of Creation the earth must have been geometrically perfect, hence spherical in shape (*11*). The original crust was then fractured by subterranean stresses, some blocks moving upward to form the continents, others sinking down to form the ocean basins. Moreover the original continents subsequently submerged and later rose from the sea. Many species of organisms have become extinct; others have changed in their characteristics through time. If large sea shells, such as the ammonites of Portland, are characteristic of tropical waters, then perhaps England at one time was in the tropical zone, he suggests. Volcanic islands have appeared in the sea at different times; perhaps one could determine their relative ages by comparing the thickness of their soils.

Time and again Hooke speaks of "past ages" or "many ages" of the earth. Referring to the slumping of rocks under attack by waves, he notes that "Most part of the Cliffs that Wall in this Island do Yearly founder and tumble into the Sea. By these means many parts are covered and rais'd by Mud and Sand that lye almost level with the Water, and others are discover'd and laid open that for many Ages have been hid." Again, when he relates his theory on the uplifting of the sea floor to form continents, he notes how by the erosion of these new lands "all those Substances which had been buried for so many Ages before, and which the devouring Teeth of Time had not consumed, may then be exposed to the Light of the Day" (*12*).

In none of these and similar statements does Hooke attach numbers to his "ages." But we should be misled if we infer that he envisioned an earth whose age is to be measured in millions or even in tens of thousands of years. For at one place where he is examining the story of the lost Atlantis, trying to determine whether it be fact or fiction, he comes to the part which places the event of subsidence at 9,000 years ago. His conclusion: "I confess that the account of the nine thousand Years is Argument enough to make the whole story to be suspected as a Fiction" (*13*).

Hooke may have been a Bohemian in his style of living; he was surely a free spirit in his physics, daring even to challenge Isaac Newton on fine points in optics and theory of gravitation; but in the marrow of his bones he was Christian. As a believer he accepted the Biblical chronology, which according to the best scholarship of the times placed Creation around 4000 B. C. Thus in trying to decipher the natural history of the earth, Hooke was constrained to work within a temporal frame of no more than 6,000 years, and he could affirm only what the Scriptures do not deny. For example, the Bible contains no reference to North America, and so Hooke can find no proof that this continent was in existence at the time of the Flood; but, he adds: "certain we are, that what was then in being was all overflowed and drowned by it, and all living Creatures, except those preserved in the Ark with *Noah*, perished by it" (*14*). At the same time, he insists, *all* the petrifactions found in the rocks cannot be accounted for by the Biblical Flood, which lasted only about 200 days—not long enough to explain the abundance of fully grown shells, nor the great thickness of sediments covering them (*15*). There must then have been other floods, caused not by rain but by sinking of the land beneath the sea (*16*).

Strong as Hooke's religious convictions seem to have been, they did not compel him to believe that one must appeal to "the immediate hand of God" to explain the changes that have taken place on the natural scene. In the case of the Israelites' passage through the Red Sea, it was the east wind that forced back the waters to leave the bottom dry, though it was the Lord that caused the wind to blow. "Nor," he continues "is there a necessity of supposing new created Causes for all the effects that we are ignorant how they are brought to pass, or to believe everything effected supernaturally, of which we cannot find out the Natural Cause." Let us first find the proximate causes, he pleads, and then proceed to the more remote (*17*).

Another Protestant tenet that colored Hooke's thinking was the proposition that the earth is in a state of decay and will soon come to an end. Martin Luther's extreme pessimism on this score had given way to a more hopeful view that the world might survive to the year 2000, but there were still those dire predictions in the Psalms, in Isaiah, and in Paul's letter to the Hebrews to the effect that "*the heavens shall vanish away like smoke, and the Earth shall wax old like a Garment*" (*18*). Hooke's commentary on the subject seeks to draw a meager comfort from the proposition that at least things *do* change in time—petrifactions, position of land and sea, and contours of the landscape—even though the changes may be from bad to worse.

> . . . we find nothing in Holy Writ that seems to argue such a constancy of Nature; but on the contrary many Expressions that denote a continual

decay, and a tendency to a final Dissolution; and this not only of Terrestrial Beings but of Celestial, even of the Sun, Moon and Stars and of the Heavens themselves. (*19*)

As evidence for this aging, Hooke pointed to the hardness of the crust, the irregularities in contours of the earth's face, and the numbers of extinct volcanos. In the beginning, he believed, the earth was made mostly of fluids, which by degrees have settled, congealed, and turned into stone, an arthritic process of stiffening similar to that seen in the aging of animals (*20*). The mountains, valleys, and basins on the face of the earth he regarded as the "Warts, Furrows, Wrinckles, and Holes of her Skin, which Age and Distempers have produced" (*21*). And, as the earth grows senile, it loses its internal energy; whatever the nature of forces that cause volcanoes to erupt and the earth to shake, these forces, like exploded powder magazines, once spent are forever dissipated. Thus, he concluded, earthquake and volcanic activities must have been more vigorous during the past than now (*22*).

In his lectures on the changes that have taken place in the earth since the beginning, Hooke touched on so many subjects and broadcast so many ideas that it is hard to do justice to all of them. His main contributions to seventeenth century historical geology lay in his proofs for the organic origin of petrifications, his proposals for extinction of species and for change within species in the course of time, and his cogent arguments supporting the proposition that subterranean forces have caused the continents to sink below the sea or rise above it.

Hooke's studies of fossils went far beyond superficial comparison of them with their modern analogues. Examining polished surfaces of petrified wood under his microscope, he could detect the same cellular structure he had discovered in cork. Working with longitudinal sections of ammonites, he could establish the basic similarities between the structure of these shells and those of the modern chambered *Nautilus*. Moreover, he drew pictures of fossils to illustrate what he was talking about. Some of these plates, published with his posthumous works, demonstrate his skills as observer and draftsman.

Hooke did not need his critics to tell him that most of his petrifications are not exactly like parts of living species. But rather than yield to the compromise that the "easy" fossils are organic and the "hard" ones sports of nature, he proposed instead that some species are extinct while others have changed their shapes in successions of generations. Granting that certain species are in fact extinct, he states:

> . . . we will grant also a supposition that several Species may really not have been created of the very Shapes they now are of, but that they have changed in great part their Shape, as well as dwindled and degenerated into a dwarfish Progeny; that this may have been so considerable, as that if we could have seen both together, we should not have judged them of the same Species. (*23*)

Hooke was stating and restating these ideas until near the end of the seventeenth century, thus long after Steno had abandoned science. Whether he or Steno should get the greater credit for the demise of the "plastic virtue" theory of petrifications is uncertain and not very important. The fact is that Plastic Virtue, after a few last whimpers, died around the turn of the century.

Postcript on Hooke

Footnotes that deal with Hooke's personality usually sketch him as a miserly, unsociable crabstick. The source of this assessment is not difficult to locate. Though no portrait of Hooke survives, Richard Waller as editor of his posthumous works provided a verbal description of the man from which any good artist could draw a likely caricature.

> As to his Person he was but despicable, being very crooked, tho' I have heard from himself, and others, that he was strait till about 16 Years of Age when he first grew awry, by frequent practicing, turning with a Turn-Lath, and the like incurvating Exercises, being but of a thin weak habit of Body, which increas'd as he grew older, so as to be very remarkable at last: This made him but low of Stature, tho' by his Limbs he shou'd have been moderately tall. He was always very pale and lean, and laterly nothing but Skin and Bone, with a meagre Aspect, his Eyes grey and full, with a sharp ingenious Look whilst younger; his nose but thin, of a moderate height and length; his mouth meanly wide, and upper Lip thin; his Chin sharp, and Forehead large; his Head of a middle size. He wore his own Hair of a dark Brown colour, very long and hanging neglected over his Face uncut and lank, which about three Year before his Death he cut off, and wore a Periwig. He went stooping and

> very fast (till his weakness a few years before his Death hindered him) having but a light Body to carry, and a great deal of Spirits and Activity, especially in his Youth. (*24*)

Waller goes on to note Hooke's irregular habits of eating and sleeping, his melancholy temper, his jealous and mistrustful nature, his monastic and penurious style of living, and his great hoard of money found locked up in a chest after his death.

With a friend like Waller, Hooke did not need enemies like Oldenburg and Newton. Were it not for his diaries, some entries in the journals of Pepys and others, and the probings of some modern historians Hooke would probably be remembered as the crouchback of Gresham College.

Hooke left at least two diaries. The earlier one shows that between 1672 and 1680 Hooke, the recluse, patronized 153 different coffee-houses and taverns in London (*25*). At such well-known meeting places as Childs, Garaways, Hercules Pillars, Joes, and Jonathans, he smoked tobacco or drank claret, sack, chocolate, port, tea, coffee, ale, beer, or white wine with his cronies, mostly fellow members of the Royal Society. Five times in 1675 Hooke (with his hair hanging down over his face??) was presented to Charles II, and on one of these occasions he presented the King with a watch of his own design (*26*). On January 1, 1676, Hooke met with members of a "New Philosophical Clubb" consisting of Fellows of the Royal Society. He was one of the organizers. His entries for that day tell something about his social habits.

In the morning he called on Tom Shortgrave, operator to the Royal Society, but missed him and so went to Garaways, a famous coffee-house in Change Alley, Cornhill, and the place where tea was first sold in London. He stayed in the Green Room there until noon, then returned to Gresham College for lunch. In the afternoon he visited three of his friends in succession, going first to the home of a Mr. Hill (probably Abraham Hill, Treasurer of the Royal Society), then to the quarters of Dr. Daniel Whistler (Professor of Geometry at Gresham), then to a Mr. Wild's lodgings. Wild and Hooke then went to Man's Coffee-House, where they met Hill, and all three went together to the home of Sir Christopher Wren. "We now began our New Philosophical Clubb, and Resolved upon Ingaging ourselves not to speak of any thing that was then reveald *sub sigillo* to any one nor to declare that we had such a meeting at all. We began our first Discourse about light upon the occasion of Mr. Newton's Late Papers. I shewd that Mr. Newton had taken my hypothesis of the puls or wave" (*27*). After Wren and Hooke had discussed the propagation of light for a while the conversation turned to the subject of water on the moon. Hooke suggested that the moon might be covered with water, reporting his "experience of seeing the bottom of the sea from the top of a high Clif that could not be seen from the top of the water. Sir Christopher affirmed no water nor River nor cloudes. Mountains not like ours nor vales." At this juncture a Mr. Wind joined the company and discoursed on ways and means of getting across swamps on floats, and of walking safely up and down icy slopes by means of spikes fastened to the heels and soles of shoes. When the Reverend William Holden arrived, the topic of discussion changed from cold to heat. "Mr. Wild told of the fellow that Kindled Tow in his mouth. Mention made of the fellow that held the Red hot iron in his teeth seen by the Royal Society." This prompted Hooke to describe his inventions of equipment to produce artificial lightning and thunder for theatrical productions that required such fireworks, which subject led to a lengthy discussion of the strength of lightning and gunpowder. Wren and Hill affirmed that no gun could fire a bullet as far as a mile. And speaking of miles, Hooke was reminded that he knew of no hill as much as a mile high. The remainder of the discussion was mostly on botany and entomology: how the juice of a certain worm will shine in the dark, whether or not plants are male and female, and how plants may be grown sealed up in glass containers. The party then went by coach to Childs Coffee-House, where in the company of John Aubrey, the antiquarian, they talked on until eleven. When Hooke got back to Gresham around midnight the gate was locked.

The several entries in Pepys' diary that mention Hooke suggest that he was an engaging conversationist, an interesting lecturer, and as we all know a fluent writer (*28*). On August 8, 1666, Pepys chanced to meet Hooke in the street and fell into conversation with him about sounds,

> and he did make me understand the nature of musicall sounds made by strings, mighty prettily; and told me that having come to a certain number of vibrations proper to make any tone, he is able to tell how many strokes a fly makes with her wings (those flies that hum in their flying) by the note that it answers to in musique during their flying. That, I suppose is a little too much refined; but his discourse in general of sound was mighty fine. (*29*)

Elsewhere in the diary Pepys records that Hooke's lecture on the recent passage of a comet was "very curious," in the sense, I take it, that it excited attention by the novel treatment of its subject matter. Another lecture of Hooke's, this one on the art of making felt, Pepys described as "very pretty." In his entry for January 21, 1665, Pepys recorded that he sat up until two o'clock reading Hooke's *Micrographia*, which he declared to be the "most ingenious book" he'd ever read (*30*).

Waller's biography does not suggest that Hooke was capable of tenderness, but the earlier diary tells a different story. For two of his three mistresses he exhibited real affection. Nell Young was one of his maids when his niece, Grace Hooke, came to live with him as a child. After Nell married and left the household, their passion matured into a lasting friendship. When Grace grew up she took Nell's place in the affections of her uncle, and her death in 1867 was a devastating experience for Hooke (*31*).

The later diary (1688–1693) shows that Hooke was still hurrying from one job to the next, when he was not enjoying companionship at the coffee-houses, or attending book auctions. His health was deteriorating, however, and two entries in 1690 tell that he was very melancholy. In 1696 he was finally awarded the money promised by Cutler. He died at the age of 67, in 1703, and it is recorded that all members of the Royal Society resident in London attended his funeral. Shortly thereafter, Isaac Newton finally accepted the presidency of the Royal Society.

CHAPTER FIVE

A Sacred Theory of the Earth

Whenever it is possible to find out the cause of what is happening, one should not have recourse to the gods.

POLYBIUS

Hooke was a bibliophile, and one of his favorite pastimes was attending book auctions. Occasionally he jotted down in his diaries the titles he had acquired and was reading. No book is mentioned more often in the later of his two diaries than Thomas Burnet's *Sacred Theory of the Earth* (*1*). Hooke's interest in this work must not be counted as yet another of his eccentricities. Shortly after the first Latin edition was published in 1681, Charles II asked Burnet to issue an English version, which appeared in 1684. In 1689 the Latin text was doubled in length, and in the following year this expanded version was printed in English. A sixth edition appeared in 1726, and *Sacred Theory* is still in print today.

No one seems to have taken a neutral position toward Burnet. He was praised by Steele, Addison, and Wordsworth, and classed by some of his distinguished admirers as the peer of Plato, Cicero, and Milton. Coleridge proposed to recast *Sacred Theory* in blank verse, but never got around to the task. Some of the *philosophes* of the Enlightenment put him on a pedestal with Descartes and Newton. On the other hand, Swift, Pope, and Gay ridiculed Burnet's ideas. All this difference of opinion aside, some things seem clear. The *Sacred Theory* sparked a war between fundamentalist Christianity and natural science that still goes on in some quarters. This contest heightened interest in geology, and cost Burnet his career as churchman (*2*).

Burnet called his theory sacred because it would justify by reason the Scriptural doctrines of Paradise and the Universal Deluge. Reason would be his first guide and if that should fall short, he would seek for light in the Scriptures. Anticipating that some critics would resent his engaging "the authority of Scripture in disputes about the Natural World, in opposition to Reason," he boldly asserted that no "truth concerning the Natural World can be an Enemy to Religion; for Truth cannot be an Enemy to Truth, God is not divided against himself" (*3*).

Beginning with the assumption that the earth was created out of chaos around 4000 B.C., Burnet reckoned that the Deluge occurred about 1,600 years later. But from where did the water of the Floods come? According to Burnet's arithmetic a body of water eight times as great as that in present oceans would be required to raise sea level above the tops of the highest mountains. Not even the heaviest torrents falling for forty days and nights could produce so much water. Nor are we to suppose that God created a superabundance of water for the Flood and then annihilated the floodwater when it had done its work. Those who call on God to make things appear and disappear "make very bold with the Deity" (*4*).

Those who would cut the knot of the problem by assuming that the Flood was only regional and not universal are denying the Mosaic account, Burnet cautioned. And if the hills of Judaea were overtopped, are we to imagine that the water stood as a great regional drop or a trembling jelly? (*5*)

Since the Scriptures tell that the earth had a beginning and will have an end, Burnet's argument continued, we cannot accept the Aristotelian idea that the earth is eternal. We have only to look about us to see that the form of the earth is changing. The higher parts tumble down during earthquakes, and wind and rain insensibly wear down the hills. "The Air alone, and the little drops of Rain have defac'd the strongest and the proudest monuments of the *Greeks* and *Romans;* and allow them but time enough, and they will of themselves beat down the Rocks into the Sea, and the Hills into the valleys" (*6*).

Granted then that the form of the earth has changed since the beginning, and granted that the earth in its present shape is incapable of sustaining a universal deluge, what must the shape of the earth have been before the Flood?

Burnet interpreted the first chapter of Genesis, which affirms that in the beginning the world was without form, to mean that the earth originated from a chaos, "a Fluid Mass, or a Mass of all sorts of little parts and particles of matter, mixt together, and floating in confusion, one with another" (7). The particles in such a mass would draw together to form a sphere, whose smooth surface would have been quite unlike the rough exterior the earth presents today. In this process of aggregation the heavier parts sank to the center of the earth and the lighter rose to the top, producing concentric shells of air and liquid around the solid core.

Thomas Burnet (Portrait from the 1734 edition of The Sacred Theory of the Earth)

We must resist the temptation to compare the three concentric shells of Burnet's model with the atmosphere, hydrosphere, and lithosphere of modern science. His "liquid shell" included not only ordinary water, but fat and oily liquids which rose to the top of that unit, so that it became divided into two parts: ordinary water below and oily liquid above. Dust settling from the primeval atmosphere was incorporated in the oily shell, converting it into "a certain slime, or fat, soft, and light Earth, spread upon the face of the Waters" (*8*). As these sedimentary particles accumulated they soaked up the oily liquor, so that the outer liquid shell became firm, its surface suitable for habitation, its base resting upon a shell of water. These waters of the "deeps" would later form the world ocean, but when first accumulated were all vaulted over with a universal layer of fertile land.

> In this smooth Earth were the first Scenes of the World, and the first Generations of Mankind; it had the beauty of Youth and blooming Nature, fresh and fruitful, and not a wrinkle, scar or fracture in all its body; no Rocks nor Mountains, no hollow Caves, nor gaping Chanels, but even and uniform all over. And the smoothness of the Earth made the face of the Heavens so too; the Air was calm and serene; none of those tumultuary motions and conflicts of vapours, which the Mountains and the Winds cause in ours: 'Twas suited to a golden Age, and to the first innocency of Nature. (*9*)

Because the axis of Burnet's paradisiacal earth was upright and not tilted, there were no seasons. In the habitable temperate zones "there was neither Heat nor Cold, Winter nor Summer—every Season was a Seed-time to Nature, and every Season an Harvest." The fertile soil, moistened by water drawn up from the oceanic shell below by the heat of the sun to settle as dew or to fall as mild showers, teemed with the vital seeds of animals as well as plants. As Eve was the mother of all mankind, so the earth was a natural incubator, the "Great Mother" of all other living creatures.

Then, as a providential act of execution upon a sinful world, the solid shell of the earth broke and fell into the shell of water below. Thus "the fountains of the Great Abysse were broken open" and the Universal Deluge overflowed all parts and regions of the broken earth. When the agitation of the Flood subsided, the waters retired to the lower places, leaving the upward-projecting fragments of the paradisiacal world standing in ruins as islands and continents.

Burnet acknowledged that the Flood was an act of God, but insisted that the divine will was served by natural causes. Providence left it to Reason to detect the causes. He conjectured that the heat of the sun caused the originally fertile shell of the earth to dry and crack, causing the waters below to boil and vaporize, and finally to explode. This cracking of the earth's shell and the flooding that followed constituted "the first great revolution of Nature," in Burnet's view.

Anticipating the denouncements of his critics, Burnet declared that "it is no detraction from the Divine Providence that the course of Nature is exact and regular." God, the maker of the celestial clock, could not only fashion a machine that would strike the hours regularly, but also design one that would fall apart at some appointed time (*10*).

Burnet thought that the world we live in is a wreck, "a broken and confus'd heap of bodies, plac'd in no order to one another, nor with any correspondency or regularity of parts" (*11*). Anyone who views the present aspect of nature in a more generous light is an "oratour," not a philosopher.

> Oratours and Philosophers treat Nature after a very different manner; Those represent her with all her graces and ornaments, and if there be anything that is not capable of that, they dissemble it, or pass it over slightly. But Philosophers view Nature with a more impartial eye, and without favour or prejudice give a just and free account, how they find all the parts of the Universe, some more, some less perfect. And as to this Earth in particular, if I was to describe it as an Oratour, I would suppose it a beautiful and regular Globe, and not only so, but that the whole Universe was made for its sake; that it was the darling and favourite of Heaven, that the sun shin'd only to give it light, to ripen its Fruit, and make fresh its Flowers; And that the great Concave of the Firmament, and all the Stars in their several Orbs, were design'd only for a spangled cabinet to keep this Jewel in. This *Idea* I would give of it as an Oratour; But a Philosopher that overheard me, would either think me in jest or very injudicious, if I took the Earth for a body so regular in it self, or so considerable, if compar'd with the rest of the Universe. This, he would say, is to make the great World like one of the Heathen Temples, a beautiful and magnificent structure, and of the richest materials, yet built only for a little brute Idol, a Dog or a Crocodile, plac'd in some corner of it. (*12*)

In Burnet's view the mountains of the earth are "nothing but great ruins" containing many hollows some of which appear at or near the surface as grottos, sea-caves, and mouths of caverns. Continued collapse of underground vaults and arches causes earthquakes. The fact that volcanic fire and smoke issue from some of these cavities indicates that "magazines of combustible materials are treasur'd up in them."

Frontispiece of Burnet's Sacred Theory. *The seven spheres portray Burnet's ideas on how the earth arrived at its present state and on what will happen to it hereafter. Beginning at the upper right and proceeding clockwise, we see in succession the primordial earth, the smooth paradisiacal earth, the earth enveloped in Noah's Flood (with the ark shown riding the waves), the earth in its ruinous present state, the earth in flames, the earth restored to a paradisiacal condition, and the earth finally transformed to a star.*

The last two books of *Sacred Theory* are more prophetic than historical. Burnet accepts the prophecies found in the second letter of Peter and in other parts of the New Testament that as the first world was destroyed by water so shall the present world be consumed by fire. As for the time when the Conflagration is to begin, Burnet could not be certain. But the flame would be volcanic; coal beds and other combustible materials would provide the fuel. And if natural agencies were not enough to liquify the whole outer shell of the earth, "let us allow *Destroying Angels* to interest themselves in the work, as the Executioners of the Divine Justice and Vengeance upon a degenerate World" (*13*). As for the *place* where the fire would be lit, Burnet the Anglican had few doubts. The territory around Rome is uniquely qualified to serve as a fuse for the Conflagration because it is at once a seat of volcanism and the seat of the Pope. We need not follow Burnet further in his account of how a new earth, better than any before, will be reconstructured from the burnt ruins of the present one.

Burnet's zealous efforts to discover in geology a "second revelation" compatible with the revelations of Scripture reflect his training and associations at the University of Cambridge. Although his date of birth is not recorded, he was probably no older than seventeen when he registered at Clare Hall in 1651. He received the bachelor's degree four years afterward, and then became first a fellow and later a proctor at Christ College. While at Cambridge he worked closely with a group called the Cambridge platonists, of which Ralph Cudworth and Henry More were prominent members.

One of the aims of these platonists was to reconcile the new science of the seventeenth century with Christian teachings. As a group they were opposed to purely mechanistic theories of nature such as the one published by Descartes in 1644. A persistent thesis in More's writings held that *all* parts of the universe proclaim "the power and providence of God." Sometimes the platonists were called "latitude men" because they valued morality above dogma and held that persons should be free to choose whatever religious organization best helped them to show their love of goodness. This tolerance extended to interpretation of Scripture. Taken literally, the sacred writings serve as guides for the unlettered; interpreted allegorically by the learned the Scriptures disclose truths more profound. For their probing beneath the surficial message of the Bible, these latitude men were sometimes called unitarians, or even atheists.

In 1671 Burnet began his travels in Europe, first as governor to the young earl of Wiltshire and later in the company of the young earl of Orrery. These travels extended over a period of three years, and in the course of them Burnet crossed the Alps. During this period he began work on his *Sacred Theory*, which was published ten years after he first left England.

In the course of his travels Burnet was overwhelmed by the irregularity of the earth's surface. He could find no pattern in the distribution of land and sea. Islands are scattered here and there "like limbs torn from the rest of the body." Promontories and capes "shoot into the Sea, and the Sinus's and Creeks on the other hand run as much into the Land; and these without any order or uniformity." Considering the whole surface of the earth, "t'is a broken and confus'd heap of bodies, plac'd in no order to one another, not with any correspondency or regularity of parts." Both the earth and the moon present the "picture of a great Ruine, and have the true aspect of a World lying in its rubbish" (*14*). Burnet the platonist, searching for a semblance of perfect geometrical form in nature, could find none. Only by looking backward in time could he envision the perfect globe of the original Creation, the Mundane Egg of Paradise with its fruitful gardens tended by men who could live hundreds of years.

Sacred Theory immediately attracted the attention of persons in high places who could read Latin. Publication of the first English edition expanded Burnet's readership and led to his professional advancement. In 1685, he was named Master of the London Charterhouse, a combination chapel, hospital, and school. Shortly afterward he became chaplain to William III. Burnet was considered a likely candidate for appointment to the Archbishopric of Canterbury.

Burnet continued to elaborate on his theory, and as he did so opposition to his views began to mount. In 1689 he issued a new and expanded version in Latin. The following year Erasmus Warren, Rector of Worlington in Suffolk, delivered a blast at Burnet in a book entitled *Geologia*. Warren attacked the sacred theory on scientific, theological, and philosophical grounds. In particular he denied Burnet's basic assumption that the present world is a great "Ruine," from which the *Sacred Theory* derived the notion that some former world must have been more perfect geometrically. To Warren the present world is a composition of surpassing beauty, with its "raised work, of Hills; the Embossings, of Mountains; the Enamelings, of lesser Seas; the Open-Work, of vast Oceans; and the Fret-work, of Rocks," not to mention "those stately Curtains overhead . . . the Clouds" (*15*). Those who might contend that mountains are not beautiful must at least concede that they are useful, Warren argued. This theme of the usefulness of mountains appeared in John Ray's influential *Three Physico-Theological Discourses*, published in 1693. Ray accepted mountains as features of a providential nature that caused rains to fall and runoff to spread replenishing soils over lowland farms.

In the course of what has been called the Burnet controversy, several alternative and more or less sacred theories of the earth were proposed. John Woodward, in his famous *Essay toward a Natural History of the Earth* (1st ed., 1695), proposed that the original crust of the earth did not collapse in ruins to cause the Universal Flood but rather was "dissolved" by the floodwaters. Later the particles settled out in order of gravity to form strata with their entombed fossils and minerals. William Whiston, who succeeded Isaac Newton at Cambridge, in 1696 accepted Woodward's idea on the origin of strata, but attributed the Flood to the close approach to the earth by a comet, which caused great rains to fall and the "Abyss" to open.

A new English edition of *Sacred Theory* appeared in 1691 and in the following year Burnet published his *Archaeologiae Philosophica*, an attempt to reconcile his theory with the account of creation in Genesis. Burnet's effort at reconciliation was widely interpreted as heterodox and profane ridicule of God's word. He was forced to resign his position at court and to retire to the Charterhouse, where he continued to write in defense of his ideas until near the time of his death in 1715.

Thomas Sprat in his early history of the Royal Society, published in 1667, anticipated that the new experimental science would be opposed by many who did not understand it. Over-zealous divines would denounce natural philosophy as the pursuit of carnal knowledge, while businessmen would dismiss it as a useless preoccupation. Misunderstandings of these kinds might be corrected by education; but ridicule of natural philosophy by the "wits" is another matter.

> I confess I believe that *New Philosophy* need not (as Caesar) fear the pale, or the melancholy, as much as the humorous, and merry: For they perhaps by making it ridiculous, becaus it is *new*, and becaus they themselves are unwilling to take pains about it, may do it more injury than all the Arguments of our severe and frowning and dogmatical *Adversaries*. (*16*)

William King, a younger contemporary of Burnet, was one of those merry people who could see the funny side of controversies involving prominent persons. King's ballad, "The Battle Royal," is a fanciful account of a debate between a trinitarian, a unitarian, and an atheist over which led the best life and had the most grace. Burnet, who is cast as the atheist, is made to proclaim:

> That all the books of Moses
> Were nothing but supposes;
> That he deserv'd rebuke, Sir,
> Who wrote the Pentateuch, Sir,
> 'Twas nothing but a sham.
>
> That as for father Adam,
> With Mrs. Eve his madam,
> And what the serpent spoke, Sir,
> 'Twas nothing but a Joke, Sir,
> And well-invented flam. (*17*)

According to King's editor, "Battle Royal" became popular and was translated into Latin and several other languages. Many of the nobility and gentry sent presents to the author, making it evident that their sentiments were against having "the mysteries of our Holy Religion discussed and canvassed after so ludicrous a manner."

Burnet failed in his diligent efforts to construct a history of the earth wherein the principal events are attributed to natural causes, without at the same time doing violence to deeper meanings found beneath the bare words of the Scriptures. By the harshest judgment, his science may seem as fantastic as his religion surely seemed heretical to many of his contemporaries—to borrow Bacon's adjectives. Nevertheless the Burnet controversy aroused interest in geology and prompted others to try casting better theories of the earth. Friends and critics alike would remember Burnet's injunction that "we must not by any means admit or imagine, that all Nature, and this great Universe, was made only for the sake of Man, the meanest of all intelligent creatures we know of: Nor that this little Planet where we sojourn for a few days, is the only habitable part of the Universe. . . ." (*18*)

In the next century this concept of the immensity of the universe in space would be coupled with the idea of the vast extension of the earth in time.

CHAPTER SIX

Telliamed's Story

All the plains which lie between the seas and the mountains were once covered by the salt waters.

LEONARDO DA VINCI

In 1692, when the publication of *Archaeologiae Philosophica* inflamed the Burnet controversy, Benoît de Maillet, French diplomat and traveller, began composing a theory which among other matters held that the earth is more than two billion years old.

Born in 1656 of a noble family of Lorraine, de Maillet's diplomatic career began with his appointment as Consul General in Egypt at age thirty-five. In 1708 he was named consul in Livorno, a post he held for seven years. Thereafter he served for five years as inspector of French establishments in the Levant and along the Barbary Coast. In 1720 he retired with royal pension from government service, and after spending two years in Paris moved to Marseilles where he died in 1738 at the age of eighty-two (*1*).

In the course of his travels, de Maillet acquired a first-hand knowledge of the geography and geology of the lands around the Mediterranean. His skill with languages, both ancient and contemporary, provided access to Arabic and Western European writings on geography, geology, and cosmology. Combining some parts of what he had read with all that he had observed, he developed a coherent and unorthodox theory of earth history in which the dimension of time would approach the order of magnitude believed to be valid by modern scientists.

To begin with, he turned away from the Mosaic account of Creation as a point of departure and adopted instead the Cartesian cos-

Benoît de Maillet (Portrait reproduced by permission of Albert V. Carozzi and the University of Illinois Press. © 1968 by the Board of Trustees of the University of Illinois Press)

mogony (*2*). Descartes had envisioned the material bodies of the universe as concentrated in a vast number of whirlpools, or vortices. Rotating in the center of each vortex is a burning star or sun with its array of opaque bodies—planets and their satellites—revolving about it. Ashes from the burning sun are propelled by the sun's rays toward the periphery of its vortex, gathering on their way dust and water which accumulate on the planets. By this process the planets grow in size, and in due course develop watery envelopes of oceans through which dust and salt continue to settle.

When the sun burns out and becomes a dark cindery body of low density, one of two things may happen. The extinguished sun may whirl to the edge of the vortex, in which case the nearest planet will take its place and flare into brightness as a new sun. Or the whole vortex may disperse as a comet and spin away to join other vortices.

According to this theory each planet is an extinguished sun. Those planets that have developed watery envelopes before their sun burned out may continue to receive water or may begin to lose water depending on their position in the new vortex.

De Maillet accepted the Cartesian idea that the earth is an extinguished sun. At some former stage of development, he theorized, the solid earth was covered by a universal ocean. Over long periods of time the ocean has been losing water by evaporation into outer space. This process of drying will continue until at last the planet will ignite and become a sun again. De Maillet estimated that about five billion years would be required for a new sun to burn out and convert to a planet, be restored in mass by addition of water and solids, and be ignited again. Presumably these cycles of burning and renovation of heavenly bodies will go on forever (*3*).

De Maillet cast his theory of the earth in the form of a dialog between two imaginary characters: a French missionary and an Indian philosopher named Telliamed (de Maillet spelled backward). His manuscript, entitled *New System on the Diminution of the Waters of the Sea,* is supposed to be the missionary's report on Telliamed's disclosures during lengthy conversations between the two at meetings in Cairo.

We soon learn that studies of the diminution of the sea have been a family project for three generations. These studies began when Telliamed's grandfather observed that near the shore bordering his seaside home a rocky prominence which had been awash during his youth had emerged above sea level years later. This phenomenon prompted grandfather to examine the rocks of mountains far inland. In the strata forming these mountains he found sea shells, sufficient proof that the sea had formerly been more extensive. Moreover, the strata were not all of a kind but were distinguishable one from another by differences in composition and color. Thus they could not have been formed all in one instant, but must have accumulated layer by layer.

So much for the strata and the fossils; how now were the mountains formed? To attack this problem grandfather invented gear for divers to explore the near-shore bottoms and map the directions of currents. Beyond that he designed a submarine for work at greater depths. The end result of these labors was the discovery that strong marine currents, moving in complicated patterns, are today heaping and molding sediment into sea mountains and valleys similar both in variety of contour and in magnitude to the mountains and valleys of the bordering lands. Thus the mountains are formed on the bottom of the sea and later emerge in essentially their original form with the diminution of the waters. *Voila!*

As for the origin and development of life Telliamed also had ready answers. When the summits of the current-built seamounts were about to emerge above the level of the diminishing ocean, the seeds of organisms began to germinate in warm shallows nearest the air. Seaweed flourished, and fish and shellfish multiplied. Remains of these organisms began to accumulate in new strata spread seaward by waves and currents. With the continued shrinking of the sea, shorelines of the emerging continents lengthened, marine life increased, and consequently the newer layers of offshore sediment became more fossiliferous. This process of emergence of the continents and building of seamounts from the sedimentary waste eroded from the shores will continue as long as there are lands to supply sediments and seas to receive them. *Soit!*

Confronted with the problem of accounting for terrestrial life, Telliamed again turned to the sea for an answer. All life, plant and animal, originated in sea water, he asserted. Fish gave rise to birds, animals that creep on the floor of the sea to animals that walk on the land, and seaweed to shrubs and trees. If flying fish marooned in reeds should transform to birds, their fins changing to wings, this would be no more amazing, Telliamed reasoned, than the transformation of a caterpillar to a butterfly. By the same token elephant seals may have been the ancestors of elephants (*4*).

Telliamed applied this same transformist theory to the origin of man. He relates mariners' tall tales of sighting sea-men and sea-women with tails like fishes but human-like from the waist up. These tritons and mermaids were as likely the descendants of seals as they are the ancestors of mankind (*5*).

How much time has elapsed since the first mountains appeared above the level of the diminishing sea? Telliamed's ingenious grandfather reasoned that knowing the height of the highest mountain and the secular rate at which sea level is lowering, an approximate answer could be given. Accordingly he constructed near his home a hydrographic station designed to measure fluctuations in sea level. Observations continued for more than 75 years indicated that sea level is falling at the rate of three inches per century (*6*). As confirmation, Telliamed cites the situation of a seaside fortress at Carthage. This edifice has basement openings evidently designed for the admission of sea water. But the base of these openings now stands five to six feet above sea level, which must have dropped by that amount in some 2,000 years (hence 3 to 3.6 inches per century). Similar situations at Alexandria and Acre supply the same results (*7*).

With the continued shrinking of the ocean, Telliamed confidently predicts, present port cities will be left high and dry as they become inland cities. By the same token, certain ancient inland settlements, or their ruins, surely originated as seaports (*8*). In terms of their present elevations, the highest of these stand perhaps 6,000 feet above sea level. Again assuming an average fall in sea level of three inches per century, the oldest maritime settlements date back some 2,400,000 years. Petrified masts of ships are found today in many inland situations, as for example in the Western Desert of Egypt—proof sufficient for the diminution of the sea (*9*).

A long period of time must have elapsed between the time the first land emerged and the time that these ancient harbors were occupied by man, Telliamed reasoned. On any reasonable assumption, he concluded, the diminution of the sea must have been going on for the past two thousand million years (*10*).

After de Maillet returned to Paris in 1720, copies of his manuscript made their rounds in French intellectual circles. The original is missing, but seven handwritten copies are preserved in French and American libraries. Struck off at various times between 1722 and 1729, these differ in detail because the author continued to revise the text until near his death in 1738 (*11*).

For editorial assistance preparatory to publication de Maillet entrusted his manuscript to the Abbé J. B. le Mascrier, who professes to have begun his work in 1732. After de Maillet's death, however, the abbot evidently had misgivings about admitting to the editorship of so heretical a book. The first edition, in French, of *Telliamed: or Conversations between an Indian philosopher and a French missionary on the diminution of the sea* did not appear until 1748. Le Mascrier's name does not appear on the title page; instead the editorship was ascribed to J. A. G. (one Jean Antoine Guer, a lawyer who apparently had no connections with de Maillet). Not until the appearance of a third edition in 1755 did the abbot acknowledge his association with the book (*12*).

Comparisons of the third edition with the extant manuscript copies have shown that le Mascrier took liberties that go far beyond the bounds of editorial privilege. His main concern was to reconcile the text with Christian orthodoxy. Certain troublesome passages he simply deleted. In some instances de Maillet's billions and millions of years were reduced by moving the decimal point three or four places to the left; elsewhere in the text quantitative estimates of geological time were translated into vague qualitative language. Glosses were introduced to make this materialistic system seem more palatable to believers (*13*).

Despite all le Mascrier's doctoring, the import of Telliamed's message was as clear to his readers as it must have been to the patient French missionary who supposedly heard the original version. Voltaire denounced *Telliamed* as the work of a charlatan who tried to play God (*14*). Other critics less well known were no more charitable. Then, as now, outraged denunciation simply worked to the advantage of booksellers. *Telliamed* became a popular book: an English translation appeared in 1750, and an American printing of 1797 capitalized on its controversial character by announcing on the title page that this is "*a very curious book.*" Not until 1968 did the authentic version of de Maillet's work appear in print, a masterpiece of scholarship by Albert Carozzi. This reconstruction, combined with Carozzi's critical commentary, shows that de Maillet's theory was not a frivolous and fanciful construction of the imagination but a serious effort at interpreting the history of the earth, mainly on the basis of observations made in the field.

Evidence that the continents have in the past been submerged beneath the sea comes from all parts of the world. The occurrence of fossilized remains of marine life in continental strata constitutes the most compelling evidence. In addition, raised beaches and wave-cut benches rise in stair-step fashion above the present strand lines in many parts of the world, as for example in Scandinavia, along stretches of the Mediterranean shore, and in California.

Thus the hypothesis of the diminution of the sea was viable in de Maillet's time. His basic assumption that the rocks forming the earth's crust were deposited from sea water identifies de Maillet as a neptunist, forerunner of a school of thought that would play a prominent and controversial role in geological developments during the interval 1775–1825 (*15*).

By applying the principle of superposition, adherents of neptunism could demonstrate that similar sequences of rocks containing distinctive kinds of fossils are present in many different parts of the

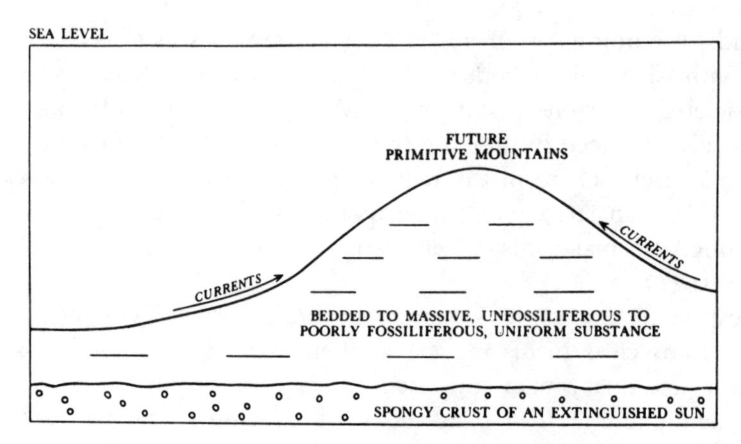

Fig. 13. Stage 1. Formation of the primitive mountains by the action of currents on the bottom of the ocean.

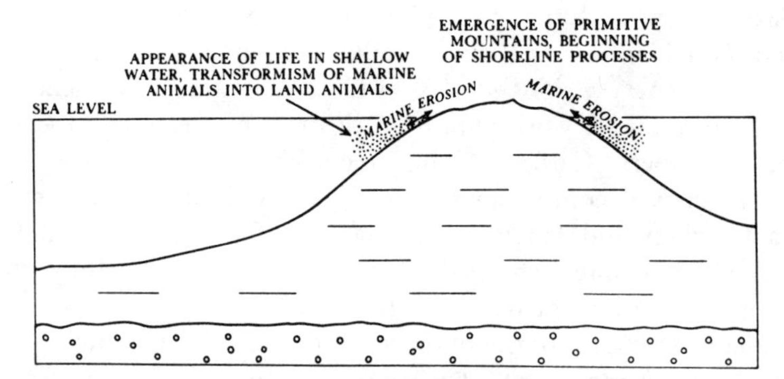

Fig. 14. Stage 2. Emergence of the primitive mountains, appearance of life in shallow waters, transformism of marine organisms into terrestrial ones, and population of the continents.

Carozzi's diagrams illustrating the evolution of the earth's crust according to de Maillet's theory (Reproduced by permission of Albert V. Carozzi and the University of Illinois Press. © 1968 by the Board of Trustees of the University of Illinois Press)

world. Two deficiencies in neptunist schemes arose from a failure to comprehend the essential difference between igneous and sedimentary rocks, and from the assumption that the earth's crust stands immobile through time. Granite and other kinds of igneous rock were regarded as sediments, either mechanically assembled or precipitated

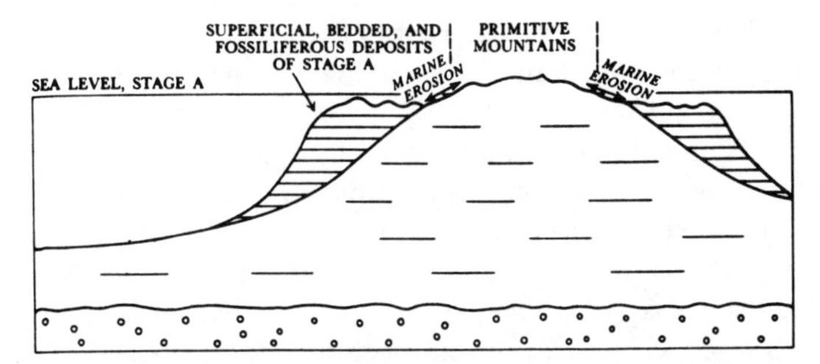

Fig. 15. Stage 3. Beginning of the formation of the secondary mountains by shoreline erosion of the primitive mountains.

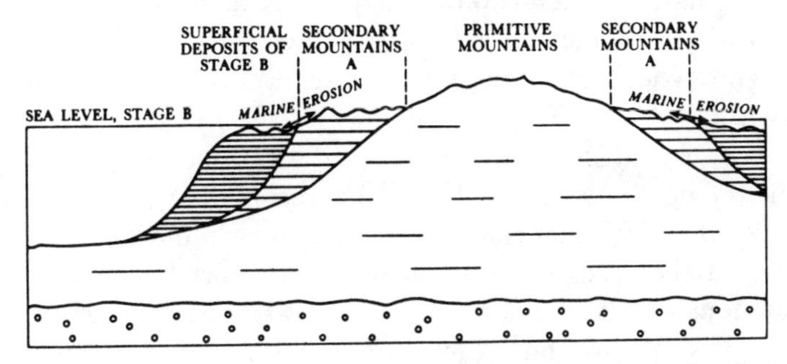

Fig. 16. Stage 4. Formation of the lower secondary mountains by shoreline erosion of the upper secondary mountains.

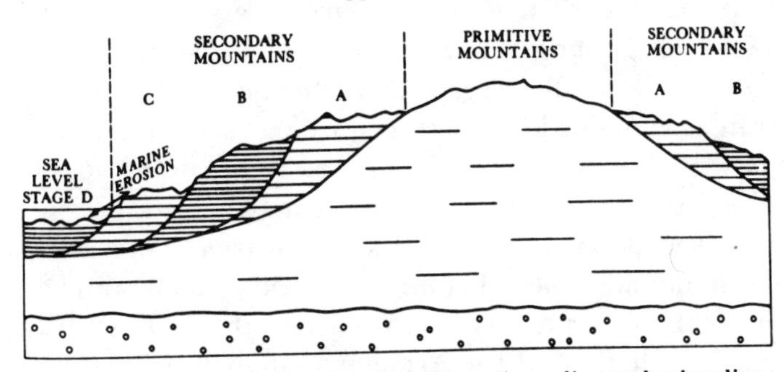

Fig. 17. Stage 5. Formation of present-day marine sediments by shoreline erosion of the lowest secondary mountains.

from a once-universal ocean. Folded strata were regarded as products of deposition over hummocky bottoms or as masses deformed by slumping of submarine sediments. Volcanic activity was generally de-emphasized and usually attributed to the burning of subterranean coal or other combustible materials. Getting rid of the waters of the shrinking ocean proved an embarrassment to neptunists of whatever stripe or vintage.

Because he denied the importance of subaerial agents of erosion—wind, streams, and glaciers—in shaping the landscape, de Maillet has been called an ultra-neptunian. Later and more sophisticated theories of the same genus allowed for the operation of these erosive agents as soon as land emerged. But as the neptunist doctrine became more refined, late in the eighteenth century, its adherents seem to have become less explicit in regard to the duration of geologic time. Emphasis on determining sequences of events from order of superposition became the rule of the day, leaving the door open to those who would encompass all these events within a Scriptural framework.

De Maillet had no patience with the tiresome and fruitless business of trying to "save the Faith" by reconciling geology with the Holy Word. His Indian sage warns that we should not measure the duration of the earth in terms of our own short lifetimes. Consider the immensity of the universe. Consider the virtually infinite number of suns, around which hundreds of millions of planets like ours revolve. Can we imagine, he asks, that all this was created in 4004 B.C.? Beyond this, is it necessary to assume that matter and motion had a beginning at all? Then may not nature be eternal (*16*)?

Nor would de Maillet accept as authentic the Biblical version of neptunism as contained in the story of Noah's Universal Flood. The mountains of Armenia are built of strata containing fossilized sea-fish, he observes. Are we to assume that the strata of these mountains were laid down in forty days as a mass of soft sediment? Or were the mountains there before the Flood and so softened by the floodwaters that sea-animals burrowed into them? Even if the Flood overtopped the highest of the Armenian mountains by 40 cubits, still it would not have covered the highest mountains of the world. Granted the Flood was an historic event, one need not assume that it was world-wide. After the ark was wrecked on the coast of Armenia, Noah and his party may be forgiven for guessing they were sole survivors of a universal catastrophe. But assuming they in fact were so, how then does one account for all the different races of man who live on the earth today (*17*)?

The Baltimore publisher who advertised *Telliamed* as "a very curious book" was surely correct. Tritons and mermaids do not exist. Venice is sinking, not rising above the sea. Not all shores have stair-step marine terraces. Petrified logs don't derive from the timbers of ancient ships. And so for many details of de Maillet's scheme, viewed from a twentieth-century perspective.

On the other hand, two grand ideas developed in *Telliamed* would continue to leaven geologic thought. The concept that terrestrial life originated in the sea brought praise from Georges Cuvier, father of comparative anatomy and vertebrate paleontology (*18*). Above all else, *Telliamed* was a geologic time-stretcher. The realization that thick sequences of strata forming high mountains were built layer by layer "from the bottom up" implied duration of time outrageously long, judged by contemporary precepts. Nevertheless, in a universe so vast as the *philosophes* of the eighteenth century were willing to concede, an earth whose age is to be measured in millions or billions of years began to seem reasonable.

CHAPTER SEVEN

The Epochs of Nature

Au commencement, Dieu créa
les cieux et la terre.

MOÏSE

Au commencement, Dieu créa
LA MATIÈRE du ciel & de la terre.

BUFFON

In 1749, a year after *Telliamed* appeared in first edition, the first three volumes of *Histoire Naturelle, Générale et Particulière* were published in Paris. The author was Georges-Louis Leclerc, then in his tenth year of service to Louis XV as Keeper of the Jardin du Roi. When appointed to this post he had been asked to catalog the collections of a natural history museum which was a part of the royal gardens. Once engaged in this curatorial work, he conceived the grand idea of preparing an encyclopedia that would treat all of nature—animal, vegetable, and mineral—in a systematic manner. Obviously too much for one man to accomplish, one would say; yet at the time of Leclerc's death 35 of a projected 50 volumes had been completed (*1*). In 1772 the author's name acquired the noble suffix *comte de Buffon*, by grace of the king. It is as Buffon and not as Leclerc that he is remembered as a prominent figure of the Enlightenment.

Buffon was born in 1707 in the Burgundian town of Montbard. His father was an officer of government whose duties included administration of the salt tax. The family suddenly became wealthy in 1714 when his mother inherited a large estate from a brother who had prospered in service to the King of Sicily. Leclerc *père* quickly proved himself equal to the task of spending his wife's money. He purchased the nearby lands and village of Buffon and a townhouse in Dijon, the capital of Burgundy. At Dijon the son attended a Jesuit college, and at the age of eighteen had completed his general studies and passed his examinations in the law (*2*).

For the next several years Buffon continued his studies privately, not in the law but in mathematics, botany, and medicine. By a fortunate circumstance, which is not documented, he met and became a friend of the young second Duke of Kingston, who was travelling in France with his tutor. Late in 1730 he embarked on The Grand Tour with these English companions, travelling through France and Switzerland and across the Alps to Rome. The Duke travelled in a style befitting his title, and through him Buffon learned some lessons in expensive living.

Buffon

Buffon's travels were briefly interrupted in the autumn of 1731 by the death of his mother. Little more than a year later his troubles increased when he learned that his father proposed to marry again, this time a woman who seemed ineligible on two counts: she was not wealthy, and she was only twenty-two years old. When he was unable to persuade his father that this union would sully the family name, he returned to Montbard to demand his share of the mother's estate. No doubt he used his knowledge of the law, which his father had urged him to acquire in the first place, to arrive at a settlement out of court, and at no greater cost than a mutual and undying alienation of affection. The settlement was handsome; in addition to a large sum in cash, he received the family real estate at Montbard and Buffon.

During the seven years that elapsed between the time of his inheritance and his appointment as keeper of the royal gardens, Buffon gained in reputation as a scientist and at the same time added to his wealth through business ventures. On his lands in Burgundy he began investigating ways of growing improved grades of timber, for use both in programs of reforestation and in building tougher ships for the French Navy. Out of this venture grew a commercial tree nursery, which Buffon later sold to the government with the understanding that he would receive a salary as manager (*3*). At the hamlet of Buffon, he built an iron foundry, where among other articles cannons for the French army and navy and iron fences for the royal gardens were cast. The design of the factory reflected the style of the builder. Blast furnace, bellows, and other gear manned by up to 400 workmen were hidden behind an elegant façade, and flanked by gardens. At a chapel in the main courtyard, services were held each Sunday (*4*). These business operations, while lucrative, did not prevent Buffon from pursuing studies of mathematics, physics, and zoology. His name was entered in the rolls of the Académie des Sciences in 1734, and five years later he was elected Associate Member in Botany. His silvaculture had paid a second dividend, this time in prestige.

In addition to the three introductory volumes of *Natural History* published in 1749, Buffon was able to complete twelve volumes on mammals, nine on birds, and five on minerals, together with six lengthy supplements. The most famous of his scientific essays, *Des Époques de la Nature*, appeared in the fifth supplementary volume, published in 1778.

In the *Époques*, Buffon attempted a synthesis of earth history from the beginning to the present, including the origin and development of life and a numerical estimate of the duration of geologic time that was outrageously long for his day (*5*).

Buffon proposed that the planets of the solar system formed as a result of a collision between a comet and the sun. The shock of impact caused about 1/650 of the sun's mass to eject as a streamer of hot liquid and gaseous matter which separated into globular masses to form the planets. The planets spun off satellites, fell back toward the sun, and began their circles around it.

Thus in the beginning the earth was a globe of molten material in the nature of a siliceous magma. The globe cooled and solidified from the surface inward. As cooling progressed the outer crust developed blisters, wrinkles, and cracks. These irregularities of surface, though negligible on a global scale, constituted a surface of high relief viewed in human perspective. Thus were formed the primitive mountains and plateaus, which are part of the main body of the earth and owe nothing to the action of water.

At one stage in the cooling of the earth the primitive siliceous crust was solid, but metallic substances with lower melting points were still liquid. Near the surface these metallic materials pooled in cavities, while at greater depths they vaporized and condensed along vertical fissures in the already solid crust. By these processes of concentration were formed the primordial ore deposits, which owe nothing to the action of water.

With further cooling a point in time was reached when water vapor could condense and fall as rain without steaming back into the atmosphere on striking the still-hot surface. Water began to gather in depressions over the earth's surface, first in the polar regions then everywhere. In time water covered the whole earth, except for local high prominences of the original crust. The work of water began.

There followed a time of intense chemical activity involving the interaction of air, earth, fire, and water. Charged with acids and salts, water infiltrated and altered the dust and slag of the primitive crust. These surficial materials disintegrated or decomposed to form deposits of sand and clay. Clay, by various degrees of hardening, changed to shale, slate, and schist. Sand, clay, and their lithified derivatives constitute one class of rocks formed by the action of water upon the raw materials of the primitive earth.

A second class of rocks made possible by water includes those which owe their origin to organisms. According to Buffon, limestones and other calcareous strata are built of the whole, broken, or pulverized hard parts of shellfish. The occurrence of limestone in the stratigraphically lower and hence older part of the stratigraphic column indicates that the universal ocean, once formed, was soon populated with innumerable shellfish.

Sediments accumulating on the bed of the ocean were swept about by currents, variously generated by the tides, by the winds, or by the rotation of the earth itself. These currents shaped the bottom sediments into relief features of the same kinds we see on the continents. Slowly the sediments hardened, and at the same time sea level fell by degrees. This retreat was due to episodic collapse of arches over the enormous caverns beneath the "blisters" of the primitive crust. With each such collapse, sea water rushed down to fill the caverns, the level of the sea fell by proportional amounts, and additional land emerged. Elevations and depressions, first shaped on the sea floor, thereby became the mountains and valleys of the land.

Vegetable life appeared on land early in the emergent stages of the continents, as evidenced by coal beds in various parts of the world. The great forests of primitive times, weakened by age and battered by winds, were sources of wood which, transported by streams and ocean currents and mixed with pyritic materials, finally became buried in the bosom of the earth. These masses of buried combustible material would burst into flame upon contact with water, giving birth to volcanos both submarine and terrestrial. Because subterranean water is necessary for their eruption, volcanos rarely exist except near the sea. With retreat of the sea, the near-shore volcanos became inactive as new ones burst into action at lower levels. Thus most volcanos in the central parts of the continents are inactive or nearly so.

At this juncture, sea level had stabilized at about its present position. The continents had emerged from the universal ocean, plastered with marine sediments molded into mountains and valleys. Over broad areas the sedimentary strata remained in their original nearly horizontal positions; but wherever the great subterranean caverns of the primitive earth had collapsed, the overlying strata were deformed. Water continued to work in its various ways. Underground water transported metallic mineral substances, concentrating them in secondary mineral deposits. Waves beat against the shores, destroying certain natural barriers such as those at Gibraltar and the Bosporus.

A last major geologic event would take place before the earth would appear in its present form. This would be the formation of the Atlantic Ocean by foundering of the lands which until this time had joined North America and Eurasia as a single great continent. Since Buffon's evidence for this event was based on paleontology, it is appropriate here to consider his views on the origin and development of life.

Buffon held that life is an inherent property of matter. The essential elements of all forms of life consist of "organic molecules." These molecules are of two kinds: animal and vegetable. Both originated by the action of heat on watery, oily, and "ductile" materials of the earth, after the surface had cooled to a critical temperature favorable to such reaction. Because these molecules formed during one stage in the secular cooling of the earth, they must be finite in number for this planet. And since other planets have passed or will pass through similar critical stages of cooling, organic molecules have been or will be produced on them also.

In Buffon's view the first organisms to appear on earth were not primitive or simple, but as complex as any living today. Due to the high temperatures prevailing at the time the organic molecules first began to form, the generative forces of nature were particularly active. Giant ammonites appeared in the ocean soon after the waters covered the earth. Mammoths and rhinoceroses were early inhabitants of the emerging continents. These large beasts were adapted to a higher temperature than their modern counterparts could endure, and so they disappeared when the temperature fell below a point they could tolerate.

Fossilized remains of mammoths are found both in Eurasia and in North America. In Buffon's reasoning this meant that the two continents were formerly joined by a land bridge which has since subsided beneath the waters of the Atlantic.

Buffon has sometimes been called a transformist, a forerunner of Darwin; but an analysis of his system shows that he was not. He maintained that terrestrial life originated in the north polar regions. These would have been the first to cool to the critical temperature required for formation of organic molecules. With continued cooling of the polar area, the plants and animals adapted to a higher range of temperature would migrate southward or else disappear. Migrating organisms would leave behind quantities of unused organic molecules. These particles would aggregate into newer organisms adapted to a more severe climate; and the new crop in turn would seek survival by southward migration as the temperature continued to fall. Thus the successive biotas spontaneously generated in the northlands were not kin to each other, only sequential and formed from the same matter.

Buffon accorded man a special place in the natural order. Taking a cue from Descartes, he distinguished man from other organisms by his gift of reason; and reason is not matter. Granted that man is the king of the earth, he is so by virtue of his intelligence and right of conquest; but he is not sovereign by divine law, he is not the purpose of Creation. Moreover life and thought are not confined to this planet alone but must be present on others throughout the vast universe.

Buffon divided the history of the earth into seven "epochs." This division is somewhat arbitrary and was not essential to his system. Evidently he hoped that the number seven would make less difficult a reconciliation of his history with the creation epic of Genesis.

Buffon's first epoch spanned the time when the earth was a globe of molten material. During the second epoch the earth consolidated, and the primordial features of relief and structure developed on its surface in the form of blisters, wrinkles, and cracks. The third epoch was the time when the ocean formed and life appeared. During the fourth epoch the waters withdrew from the continents and volcanos became active. Elephants and other kinds of animals which at present live in warm climates populated the northern parts of the continents during the fifth epoch. The major physical event of the sixth epoch was the separation of the American and Eurasian continents. The seventh epoch is the Age of Man, the time when man acquired dominion over nature in the senses of understanding her workings, exploiting natural resources, and in various ways modifying the natural scene.

The problem now was to estimate the time required for the earth to cool from its original state of fusion to its present temperature. Having thus calculated the age of the earth, the duration of the several epochs could be estimated as fractions of this total.

In his search for numbers to measure the age of the earth in years, Buffon set up an experiment in his iron foundry. He had his workmen fashion ten balls of iron graduated in diameter by half-inches up to a maximum of five inches. These balls were heated to near the melting point. Then he measured the time required for each ball to cool, first to the point that it could be touched without burning the fingers, and then to the point that the temperature was the same as that of the air in a nearby cave. These experiments showed that with each increase of a half-inch in diameter, the time required for cooling to the first point increased by 12 minutes, and for cooling to cave temperature by 54 minutes. With a bold extrapolation he calculated that a globe of molten iron the size of the earth would require 49,964 years and 221 days to cool to the point that it would not burn the hand when touched; and 96,670 years and 132 days for the temperature to fall to the present temperature of the earth. In other calculations Buffon arrived at a figure of 4,026 years for a globe of the same size to pass from a molten to a consolidated condition.

Buffon repeated the experiments in cooling, this time using mixtures of metallic and non-metallic substances more like the actual composition of the earth. He made corrections in his figures to allow for retardation in cooling due to the heat that the earth would receive from the sun. To follow him through all of his arithmetic would be tiresome: the key figures he finally published are the following:

2,936 years for the earth to consolidate from a molten condition;
37,500 years for cooling before the earth could be safely touched; and
75,000 years for the earth to cool to its present temperature.

These figures are not cumulative: according to Buffon the earth was 75,000 years old in the 18th century A.D. Interpolating between the above figures, he estimated that the oceans formed and life appeared when the earth was about 35,000 years old, that the stabilization of sea level took place at between 50,000 and 55,000 years, that the mammoths appeared in Siberia around year 60,000, that the continents separated around year 65,000, and that the southward migration of elephants ended at year 70,000 (*6*).

This published chronicle was actually Buffon's "short version." In manuscripts which were not publicized until the nineteenth century, he opted for a much longer scale of time. Among other considerations he was impressed with the length of time that must have been required for thick sequences of limestone strata to form. Not centuries, he conjectured, but centuries of centuries must be required for the growth of shellfish, their accumulation as entire shells or fragments in successive strata, and their hardening to form the rocks we see in our quarries. Buffon estimated that the rate of deposition of mud in the seas of today averages less than five inches a year. Thus a single formation of clay 6,000 feet thick would have required more than 14,000 years to accumulate. Buffon was no field geologist, but he must certainly have been impressed by the enormous thickness of sedimentary rocks he saw when crossing the Alps on his grand tour.

Of the several "long chronologies" found in the unpublished manuscripts, the longest one estimates the age of the earth at nearly three million years, and would have life appear after the planet was between 700,000 and a million years old (7). Internal evidence strongly suggests that Buffon favored the longer version. Why, then, did he elect to publish the shorter? Surely not to appease the theologians, who would have considered the proposed 75,000 years for the age of the earth as much a departure from Scriptural truth as 3,000,000. The best guess is that he thought the lower figure was about as much as the traffic would bear. His readers simply would not be prepared to contemplate *le sombre abîme* of time plumbed in millions of years.

Natural History made Buffon famous, not only as a scientist but also as a literary figure. His election to the Académie Française in 1753 came as recognition of his accomplishments as stylist, and his *Essay on Style* is still highly regarded. "Style is the man himself," Buffon declared in a phrase often quoted. The reference of course is to literature, but in the case of this man the meaning could also apply to style of living. Though already an international figure in his middle years, and wealthy too, he continued to be self-disciplined and industrious.

In 1767 Buffon acquired financial control of his encyclopedia and so became his own publisher. This must have been a sound investment, because *Natural History* became a popular work during and after his lifetime. One can count fifty-two complete editions in French, and a total of twenty others translated into German, Italian, English, Spanish, and Dutch (*8*).

Buffon was less successful in his family affairs. When he was forty-five years old he followed his father's example by marrying a woman of twenty who brought no dowry with her. For his only son, born in 1764, he provided the best in training and polite upbringing that money and social connections could afford. Unfortunately, the boy was not equipped to carry on the intellectual tradition of the father. "Buffonet," as he was nicknamed, has been described as a crackbrain and rakehell. He did manage to marry into high society, but his wife had a scandalous affair with the Duke of Orleans. At age 30 he was guillotined, a victim of the French Revolution.

In 1788 Buffon died at his apartment in Paris. Buffon once observed that the scientific study of nature required of the investigator two opposite qualities of mind: the ability to conceive the whole grand picture at a glance, and the capability of focusing patiently upon a single aspect of the whole (*9*). Most certainly he qualified on the first count: his vision of the history of nature was panoramic, and he conveyed it with consummate skill and grace.

As for focusing upon a single aspect of the whole, he was content to glean most of his facts from the literature and from personal correspondence. Let others provide the bricks; he would design and build the edifice! Officials of the French colonies supplied information on remote parts of the world. Others of his better known correspondents included Catherine the Great of Russia, the Kings of Denmark, Poland, Sweden, and Prussia, Benjamin Franklin, and Thomas Jefferson (*10*).

Buffon had a serene confidence in the power of scientific reasoning. His basic assumption was that nature is rational, product of a rational Creator. Hence nature can be understood. He could foresee no limits to human understanding of the universe. We begin our interrogation of nature, he argued, by comparing like things and events. Anything unique, unrelated to anything else, is beyond comprehension. Thus God is incomprehensible, because unique. Appealing to God to explain natural phenomena is a form of mental laziness, as pretending to understand God's designs is a symptom of human weakness. Science must deal solely with secondary causes, never with those primary causes which are God's will. Hence science is concerned with the *how* of secondary causes, not with the divine *why*.

In attempting to establish the sequence of changes the earth has undergone since its beginning, Buffon held that the scientist should appeal to those ordinary causes of change that are in operation today (*11*). Invoking extraordinary or catastrophic causes leads to the invention of groundless hypotheses, he warned. He dismissed as wild and romantic speculations Woodward's Deluge which was supposed to have dissolved all the rocks of the earth's crust, and Whiston's comet which opened the floodgates of the Abyss.

The just reward of the scientist's labors, Buffon concluded, is to perceive the simplicity and economy of means by which natural forces have shaped the earth. The basic operations of nature, he believed, are few in number, involving force, motion, chemical reaction, heat, and vital energy. He placed great emphasis on force as an underlying cause of change. Force, he reasons, is the result of the mutual attraction between bodies which produces motion; chemical reaction the result of attractions between the smaller particles of matter; and life itself the result of attractions of small particles combining under the influence of heat to form organic molecules.

These sentiments were displeasing to the Faculty of Theology at the University of Paris. In 1751, two years after the first volumes of *Natural History* appeared, Buffon escaped censure at the hands of the theologians only by dismissing his theory for the origin of the planets as "pure philosophical speculation" (*12*). In writing the *Époques* Buffon tried to forestall any additional threat of censure by explaining how his theory was not actually contrary to Christian dogma. His exegesis is all one would expect of genius artfully at work on an insoluble problem (*13*). The first verse of Genesis, he argued, should be interpreted to read: "In the beginning God created *the materials* of Heaven and Earth": all else would follow according to the laws governing matter and energy. The theologians were not pleased, but their investigations stopped short of censure, apparently at the suggestion of the King (*14*).

Through his writings Buffon aroused popular interest in natural science. Beyond that he was the first to undertake geophysical experiments in the search for numbers to measure the age of the earth. In so doing he broke with the prevailing Biblical chronology which compressed earth history within a time frame of a few millennia, and offered instead a chronicle calibrated in tens and scores of thousands of years. In this respect the *Époques* were epochal. So was the author, at least in the view of one admirer.

> Qu'il soit béni le jour qui vit naître Buffon!
> Buffon sera, chez la race future,
> Pour les amis du vrai, du beau, de la raison
> Une époque de la nature. (*15*)
>
> GUÉNEAU DE MONTEBEILLARD, 1778

CHAPTER EIGHT

View from the Brink

> The abyss from which the man of science should recoil is that of ignorance . . . .
>
> JAMES HUTTON

In 1788, the year of Buffon's death, James Hutton's *Theory of the Earth* appeared in Volume I of the *Transactions of the Royal Society of Edinburgh.* The ideas expressed in that long essay, together with the author's later elaborations of them, are considered by many to mark the beginning of modern geology. Hutton emphasized the work of streams in shaping the surface of the land. He recognized the significance of angular unconformities between two sequences of strata as recording successive events of deposition, deformation, erosion, and burial. Through his own field observations he established the igneous origin of granite and basalt. In constructing his theory he appealed to processes currently in action, as causes sufficient to account for past changes on the earth. Thus he held that most valleys have been formed through protracted erosion by the streams that now occupy them, not by some catastrophic flood. He proposed that in the past whole continents have been leveled by the day-to-day action of waves and running water. The time required for this decay of the continents would necessarily be vast; but this was of no concern to Hutton, who steadfastly believed that geologic time is virtually infinite in duration.

Any young scholar who suffers pangs of anxiety concerning the choice of a suitable career should take heart from the story of James Hutton, who changed his course three times and was sixty-two years old when his first major work was published.

Hutton was born in Edinburgh on June 3, 1726. His father, a merchant who had for several years also served as City Treasurer, died when the boy was young. That Hutton received a liberal education seems to have been due to the influence of his mother. In 1740 he entered the University of Edinburgh as a student of the humanities. However, he soon became interested in chemistry, and largely on his own initiative acquired some proficiency in that branch of science.

Upon graduation he entered upon a career in business, as apprentice to a legal firm. This employment lasted less than a year, for reasons his biographer makes clear.

> The young man's propensity to study continued, and he was often found amusing himself and his fellow apprentices with chemical experiments, when he should have been copying papers, or studying the forms of legal proceedings; so that Mr. CHALMERS soon perceived that the business of a writer was not that in which he was destined to succeed. With much good sense and kindness, therefore, he advised him to think of some employment better suited to his turn of mind, and released him from the obligations which he had come under as his apprentice. (*1*)

In 1744 Hutton enrolled again at the University of Edinburgh, this time as a pre-medical student. After three years he went to Paris for additional training in anatomy and chemistry, then on to Holland for a terminal year of study. He was 23 years old when he received the M.D. degree from the University of Leiden.

Late in 1749 he moved to London still unsettled on the choice of a career. Prospects of establishing a profitable practice as physician in Edinburgh seemed dim. Meanwhile, however, he had been relieved of financial worries by income from a business venture with a friend in the manufacture of sal ammoniac from chimney soot. Also he owned by inheritance a farm near Duns in the southern uplands of Scotland. In 1750 he decided to become a farmer.

> As he was never disposed to do anything by halves, he determined to study rural economy in the school which was then reckoned the best. . . . He went into Norfolk, and fixed his residence for some time in that country, living in the house of a farmer, who served both for his landlord and his instructor. . . . He appears, indeed, to have enjoyed this situation very much: the simple and plain character of the society with which he mingled, suited well with his own, and the peasants of Norfolk would

> find nothing in the stranger to set them at a distance from him. . . . It was always true of Dr. HUTTON, that to an ordinary man he appeared to be an ordinary man. . . .(*2*)

While based in Norfolk, Hutton frequently made excursions by foot into other parts of England. His purpose was to learn how agriculture is practiced under different situations. In the course of these journeys he became "very fond of studying the surface of the earth," as he noted in a letter written in 1753 (*3*). In 1754, after touring the Low Countries of Europe to learn more about farming, he returned to Scotland and settled on his rural estate. There, with the aid of a ploughman recruited from Norfolk, he "set the first example of good tillage which had been seen in that district." Hutton is also credited as being the first farmer in Scotland to use the two-horse plough (*4*).

Dr. Hutton lived on his farm for about fourteen years, occasionally making trips to Edinburgh and geological excursions into various parts of Scotland. All the while he had been developing his theory of the earth, the main ideas of which seemed to have crystallized in his mind around 1760. He retired from farming in 1768, moved to Edinburgh, and resided there the rest of his life.

In Edinburgh Hutton readily found his way into the intellectual and social circles of that prospering city. Among his close associates were Sir James Hall, the founder of experimental geology; Adam Smith, author of *Wealth of Nations;* Dr. Joseph Black, the chemist who discovered carbon dioxide; Robert Adam, the architect and designer of furniture; John Playfair, mathematician and physicist; and James Watt, who invented the steam engine.

With these friends Hutton was content to discuss and debate his theory of the earth, without evident desire to publish his ideas; "for he was one of those who are much more delighted with the contemplation of truth, than with the praise of having discovered it" (*5*). But in 1783, when the Royal Society of Edinburgh received its charter, his zeal for supporting an institution "which he thought of importance to the progress of science in his own country induced him to come forward and to communicate. . . . a concise account of his theory of the earth" (*6*). This concise account is Hutton's essay of 1788 (*7*). Hutton's reasoning in this essay runs as follows.

The earth functions as a machine whose purpose is to sustain life. This machine has four parts: a central body which is concealed from our view, a body of water which we call the ocean, an irregular and discontinuous body of land rising out of the ocean, and a body of atmosphere enveloping the entire globe.

In addition to providing a home for a host of organisms, the ocean serves as "the receptacle of our rivers and the fountain of our vapours."

The atmosphere makes fire possible, and provides the breath of life for animals. Also the atmosphere is an agent of circulation, "raising up the water of the ocean and pouring it forth upon the surface of the land."

The land sustains animal and vegetable life with the aid of water derived ultimately from the ocean.

Decay is a necessary function in a habitable world. For animals require plant food; plants require soil; and soil forms by the decay of rocks.

Decay is also seen in the wearing down of the land by running water. "Our fertile plains are formed from the ruins of the mountains." Ultimately the streams carry soil and other earth materials to the shore, where waves and currents transport this waste of the land to the bottom of the ocean.

Therefore in the course of time, which to nature is endless, "we may perceive an end to this beautiful machine." With the wasting away of the continents by erosion, the day may come when the earth will no longer be capable of sustaining man and other terrestrial life.

Unless we can discover some built-in mechanism by which this machine is naturally repaired and rescued from decay, we must draw one of two conclusions: "the system of this earth has either been intentionally made imperfect, or has not been the work of infinite power and wisdom" (*8*).

At this point Hutton has come midway in developing a cyclic theory of the earth, involving alternating destruction and renovation of the continents. He proceeds next to describe the constructive phase of the cycle.

Most of that part of the world which is now land is made of strata originally deposited at the bottom of the sea. The evidence provided by the remains of marine organisms in these strata admits of no other conclusion.

Thus we are faced with two questions. How could "such continents as we actually have upon the globe. . . . be erected above the level of the sea"? And, how did the strata on the continents acquire their present hardness and solidity?

James Hutton (left) and John Playfair. Photograph of busts at Edinburgh by Donald B. McIntyre, and reproduced here with his permission.

Surely these strata, when first deposited, consisted of particles that were loosely aggregated. Water and heat are the two agencies to which we might appeal for the consolidation of rock from sediment.

The agency of water can be eliminated on several counts (*9*). If we assume that the spaces between the sedimentary particles "first occupied by water . . . are afterwards to be filled with a hard and solid substance," how did the cementing material get into the water in the first place, and what happened to the water at the end of the process? Also, certain strata are cemented with materials, such as siliceous and sulphurous substances, that are not known to be soluble in water.

That leaves heat. Though we do not know the cause of the earth's internal heat, we can visualize its effects on strata. Heat could bring strata to a state of fusion from which they would later congeal to become the hard bodies we see today. The downward pressure exerted by the ocean and by overlying strata would assist in the work of compaction. During the baking process, foreign matter could be introduced into openings either as molten materials, or as deposits from hot vapors. For example, the flint nodules which are so common in the chalk of England must have congealed from a state of fusion after their injection into the rock as droplets of molten material (*10*). A wealth of evidence leads to the conclusion that "all the solid strata of the world have been condensed by means of heat and hardened from a state of fusion" (*11*).

Hutton now turns to the problem of explaining how the hardened strata at the bottom of the sea are raised to form new land. Again he appeals to heat, this time to the power of heat to make bodies expand.

When originally deposited on the sea floor, the strata now forming the continents must have been horizontal or nearly so. But in many places these strata are fractured, dislocated, and contorted. In such situations, metallic mineral deposits commonly form veins filling fractures; these substances must have originated in "the bowels of the earth, the place of power and expansion." The mineral veins occupy zones of weakness produced by "violent fracture and unlimited dislocation." These columns of fluid metals and stone were forced up from the bottom of the ocean to heights equaling those of the Andes and Alps. To appreciate the expansive power of subterranean heat,

E-W section, Northern Granite, Isle of Arran, Strathclyde, Scotland. Granite of Tertiary age (forming the higher parts of the mountain in the center) has intruded and uparched strata of Precambrian and younger age. The granite and adjacent strata have in turn been cut by dikes. One of the several situations Hutton offered as proof that granite is an intrusive igneous rock. (From James Hutton's Theory of the Earth: the lost drawings, *published by the Scottish Academic Press, and reproduced here with the permission of Sir John Clerk of Penicuik, Bart.)*

we need only to see it in action. During an eruption of Mt. Etna, for example, a column of molten rock is forced upward from a source somewhere below sea level to points of discharge at immense heights above.

This raises the question as to whether this internal heat has operated always in the past and everywhere over the globe. The answer would seem to be yes, on both counts. Ancient lavas, extruded from volcanos long extinct, are widespread. Moreover there are bodies of "unerupted lava" that cut across the strata or follow partings between them. These intruded masses are common in the British Isles and Western Europe.

Volcanic activity of the kind we see at Etna is thus only the surficial manifestation of violent subterranean heat. Despite their occasional destructiveness, volcanos have a benign purpose.

> A volcano is not made on purpose to frighten superstitious people into fits of piety and devotion, nor to overwhelm devoted cities with destruction; a volcano should be considered as a spiracle to the subterranean furnace, in order to prevent the unnecessary elevation of land, and fatal effects of earthquakes.(*12*)

In other words, volcanos are safety valves in a heat machine.

Finally Hutton considered the thorny question of the earth's age. The present continents, he reminds us, are formed from the waste of older continents. The present lands themselves are eroding away, and the sediments shed from them are being stored on the ocean floor as the foundations of continents that will someday rise above the sea.

If one could measure the annual rate at which the lands are being destroyed by erosion, this figure could be the basis for estimating in years the time required for completion of the destructive phase in the cyclic operation of the earth machine. But if we go to the shore to measure the rate at which waves are eating away at the land, we find the retreat of the land so gradual that its rate can't be measured in a year or a lifetime. Or if we turn to the geographic records of the Romans and Greeks, their accounts "give no measure of a decrease, or are not accurate enough for such a purpose." The passage between Sicily and Italy appears to be no wider now than in Roman times, and the Isthmus of Corinth is apparently the same as it was more than a thousand years ago. Thus the wasting of the continents by the sea is a process so slow that an indefinite but undoubtedly vast amount of time will be required for the destruction of the continents.

In closing his essay, Hutton concedes that we do not have all the information about the earth that one might wish.

> But we have got enough; we have the satisfaction to find, that in nature there is wisdom, system, and consistency. For having in the natural history of this earth, seen a succession of worlds, we may from this conclude that there is a system in nature. . . . But if the succession of worlds is established in the system of nature, it is in vain to look for anything higher in the origin of the earth. The result, therefore, of this present enquiry is, that we find no vestige of a beginning,—no prospect of an end. (*13*)

The publication of Hutton's theory in 1788 produced no immediate reaction. It was simply ignored. Perhaps readers were weary of theories of the earth; perhaps Hutton's literary style was too opaque. The best guess, however, is that Hutton had gone against the grain of contemporary geologic thought. Neptunism was then the prevailing doctrine.

In his essay Hutton had little to say about theories of others, except to dismiss them all as unreasonable. Of Burnet's *Sacred Theory*, he noted: "This surely cannot be considered in any other light than a dream, formed upon the poetic fiction of a golden age, and that of iron which had succeeded it. . . ." In *Telliamed*, he conceded, "we have a very ingenious theory . . . which has something in it like a regular system, such as we might expect to find in nature; but, it is only a physical romance, and cannot be considered in a serious view . . ." (*14*). Buffon, on the other hand, had offered a theory "not founded on any regular system, but upon an irregularity of nature, or an accident supposed to have happened to the sun." We are not to ascribe the production of the "beautiful system of this world" to some accident such as "the error of a comet" (*15*).

Hutton noted that all neptunist theories, in whatever format, would have the seas retreat rather than have the continents rise. But, he argued, this would require "sinking the body of the former land into the solid globe, so as to swallow up the greater part of the ocean after it," which "if not a natural impossibility" is "at least a superfluous exertion of the power of nature." The land naturally decays: why then "employ so extraordinary a power, in order to hide a former continent of land, and puzzle man" (*16*)?

James Kirwan of Edinburgh, chemist, mineralogist, and leader of the neptunist school in Ireland, took strong exception to these remarks. In 1794 he delivered a blast at Hutton's theory, noting not only that it was contrary to Mosaic history but also that it plunges the reader into an abyss of time "from which human reason recoils" (*17*). Hutton, who was recovering from a serious illness when Kirwan's paper appeared, quickly regained his spirits and published an expanded *Theory of the Earth* in 1795. In that work of two volumes and more than 1,100 pages, he reaffirmed his views and replied to the criticisms of Kirwan and others.

Kirwan's attack focused upon five issues: the erosion of soil by running water, the relative abundance of stratified rocks formed from unconsolidated sediments on the floor of the sea, the source of the energy for subterranean heat, the origin of granite, and the duration of geologic time (*18*).

Kirwan charged that Hutton would have soil constantly washed into the sea. Hutton denied this was his meaning: soils are *necessarily* and episodically washed to the sea in the fullness of time. Kirwan challenged Hutton's view that most strata of the earth originally accumulated as loose sediment on the sea bottom, citing successions of slate, marble, and other types of rocks that contain no fossils indicating marine origin. Hutton countered by claiming that one should not expect to find fossils in *every* stone. Over the long periods of time elapsed since the strata were formed, fossils must commonly be obliterated by mechanical or chemical alterations.

Kirwan interpreted Hutton's "subterranean heat" to be produced by the burning of coal, sulphur, and bituminous substances deep in the earth. Hutton hotly denies this: he does not know the source of the heat, but recent volcanic eruptions and ancient masses of basalt tell us that the heat is there now and has been for a long time.

With regard to the earth's age, Kirwan took Hutton's "no vestige of a beginning" to mean that this world has existed throughout all eternity. Hutton wondered if now he was to be charged with atheism. To say that we *see* no vestige of a beginning, he insisted, is only a confession of human limitations, not a claim that "those things which we see have always been, or been without a cause. . . . But in thus tracing back the natural operations which have succeeded each other, and mark to us the course of time past, we come to a period in which we cannot see any farther" (*19*). At that point we have reached "the limit of our retrospective view of those operations which have come to pass in time, and have been conducted by supreme intelligence."

Hutton died at Edinburgh in 1797, leaving in manuscript form the copy for a third volume of his *Theory*, which was later discovered and published one hundred and four years after the first two volumes appeared. Considering that all of his publications were issued during the last twenty of his seventy-one years, his productivity as a writer is amazing. In addition to the essay and two volumes on his theory of the earth, he wrote papers on the classification of coal, on the theory of rain, on linguistics, and on the origin of granite. He is the author of a one-volume treatise on physics, and a three-volume work on the principles of knowledge. When he died he left a manuscript on the principles of agriculture long enough for two volumes. This output of publications and manuscript is explained by the fact that when Hutton finally settled in Edinburgh, he could spend his time as he chose, and he chose to spend most of it in research and writing.

> Indeed his manner of life, at least after he left off the occupations of husbandry, gave him such a command of his time, as is enjoyed by very few. Though he used to rise late, he began immediately to study, and generally continued busy till dinner. He dined early, almost always at home, and passed very little time at table; for he ate sparingly, and drank no wine. After dinner he resumed his studies, or, if the weather was fine, walked for two or three hours. . . . The evening he always spent in the society of his friends. (*20*)

On the testimony of Playfair, Hutton was no dour Scot at these evening gatherings.

> A brighter tint of gaiety and chearfulness spread itself over every countenance when the Doctor entered the room; and the philosopher who had just descended from the sublimest speculations on metaphysics, or risen from the deepest researches of geology, seated himself at the tea-table, as much disengaged from thought, as chearful and gay, as the youngest of the company. (*21*)

Hutton was no less animated when he made or learned of new discoveries in almost any field of knowledge.

> This sensibility to intellectual pleasure, was not confined to a few objects, nor to the sciences which he particularly cultivated: he would rejoice over WATT's improvements of the steam engine, or COOK's discoveries in the South Sea, with all the warmth of a man who was to share in

the honour or the profit about to accrue from them. The fire of his expression, on such occasions, and the animation of his countenance and manner, are not to be described; they were always seen with great delight by those who could enter into his sentiments, and often with great astonishment by those who could not. (*22*)

Some of those who could not enter into Hutton's sentiments included the guides who once led him on a field trip along the valley of the Tilt, a small stream in the Grampian Mountains. Hutton was looking there for evidence to support his idea that granite is of igneous rather than sedimentary origin. When the party found dikes of red granite cutting the black country rock sparkling with mica he was overcome with excitement and joy.

The sight of objects which verified at once so many important conclusions in his system, filled him with delight; and as his feelings, on such occasions, were always strongly expressed, the guides who accompanied him were convinced that it must be nothing less than the discovery of a vein of silver or gold, that could call forth such strong marks of joy and exultation. (*23*)

Hutton was always concerned more with the *kinds and causes* of changes that affect the earth than with ordering these changes in some chronologic frame. He left to the neptunists the tedious task of arranging strata in a series according to their relative ages. Rocks, minerals, fossils, soils, lava, water, air, and organisms—the material parts of nature—all were subjects of his investigations, but mainly as sources of information on the changes by which they have come to be what they are. In this world change is inevitable because everything moves.

Every material being exists in motion, every immaterial being in action and in passion; rest exists not any where. . . . Surely it is contrary to every species of philosophy . . . to found a system on the inutility of repose, or place perfection in the vacuity of rest; . . . when every real information which we have is derived from a change. . . . (*24*)

We have only to look about us to see that change is always taking place, as

. . . .from the top of the mountain to the shore of the sea. . . . every thing is in a state of change; the rock and solid strata dissolving, breaking, and decomposing, for the purpose of becoming soil; the soil travelling along the surface of the earth, in its way to the shore; and the shore wearing and wasting by the agitation of the sea. . . . (*25*)

With regard to the magnitude of effects produced by individual changes, Hutton allowed great latitude. The effects of stream erosion upon the configuration of a continent would hardly be perceptible after days, years, or even centuries; but the end result in a sequence of cumulative day-to-day erosional events over an indefinitely long period of time would be the destruction of the continent. On the other hand, single events could be the causes of profound changes. As Hutton theorized, "there is a more sudden destruction that may be supposed to happen sometimes to our continents. . . ." If the lands are raised above the sea by the expansion of matter, it follows that the expanded matter must become more "rarified" (that is, less dense) than before. "We may thus consider our land as placed upon pillars, which may break, and thus restore the ancient situation of things when this land had been originally collected at the bottom of the ocean" (*26*). The uplift of the continents, no less than their foundering, Hutton thought might be catastrophic, as evidenced in the "violent fracture and unlimited dislocation" of the uplifted strata.

Although Hutton admitted violent and destructive causes to this system, he would have none of supernatural origin, except the "first cause" of Creation. He took no stock in the tradition of Noah's Flood. Surely, he observed, "general deluges form no part of the theory of the earth; for, the purpose of this earth is evidently to maintain vegetable and animal life, and not to destroy them" (*27*). He insisted that changes we witness today, or infer to have happened in the past, must be attributed to natural causes; otherwise there can be no science. In his words, "if the stone . . . which fell today were to rise again tomorrow, there would be an end to natural philosophy, our principles would fail . . ." (*28*).

Hutton drew two analogies as models to explain the workings of the natural world: one mechanical, the other organic. Mechanically the world is like a machine powered by heat. He expresses this idea time and again in unfolding this theory. Less frequently he compares the earth to an organism. In the context of the "wasting of the land and the necessity of its renovation," he speaks of the "physiology of this earth," as though the world were itself an organism. Changes in the physical state of the heat machine are compared with the metabolic changes in organisms. "This earth, like the body of an animal, is wasted at the same time it is repaired. It has a state of growth and augmentation; it has another state, which is that of diminution and decay." (*29*)

Machine or organism, the earth has a purpose: to sustain life in the first instance or to live on in the second. In either case we are led to see "a beautiful economy in the works of nature." And, "what a comfort to man, for whom that system was contrived, as the only living being on this earth who can perceive it; what a comfort, I say, to think that the Author of our existence has given such evident marks of his good-will towards man" (*30*). Hutton was no atheist: he was a deist and providentialist.

Hutton declined to speculate on how the earth began or how it may end. He attempted no numerical estimate of time past, based upon the secular cooling of the earth, upon the diminution of the sea, or upon the rate at which sediment drops into the hourglass of the ocean. In his view geologic time must be immeasurably vast to accommodate the succession of events which he inferred to have taken place. He appears to have been the first to comprehend the implications with regard to time as expressed in what geologists now call angular unconformities separating two thick sequences of strata. Which takes us to Siccar Point.

In the seacliffs at Siccar Point, a little north of St. Abbs on the rocky eastern coast of the southern uplands of Scotland, almost horizontal strata of the Old Red Sandstone rest upon highly contorted strata of a system which we now call Silurian and which Hutton called "the schistus." Hutton's account of his visit to the Point is prosaic.

> It was late in the spring 1788 when Sir James (Hall) left town, and Mr. Playfair and I went to Dunglass about the beginning of June. . . . Having taken boat at Dunglass burn we set out to explore the coast. . . . We found the junction of that schistus with the red sand-stone. . . . at St. Helens. But, at Siccar Point, we found a beautiful picture of this junction washed bare by the sea. The sand-stone strata. . . . are partly remaining upon the ends of the vertical schistus; and, in many places, points of the schistus strata are seen standing up through among the sand-stone. . . . Behind this again we have a natural section of those sand-stone strata, containing fragments of the schistus. . . . We returned perfectly satisfied; and Sir James Hall is to pursue this subject farther when he shall be in these mountains shooting muir game. (*31*)

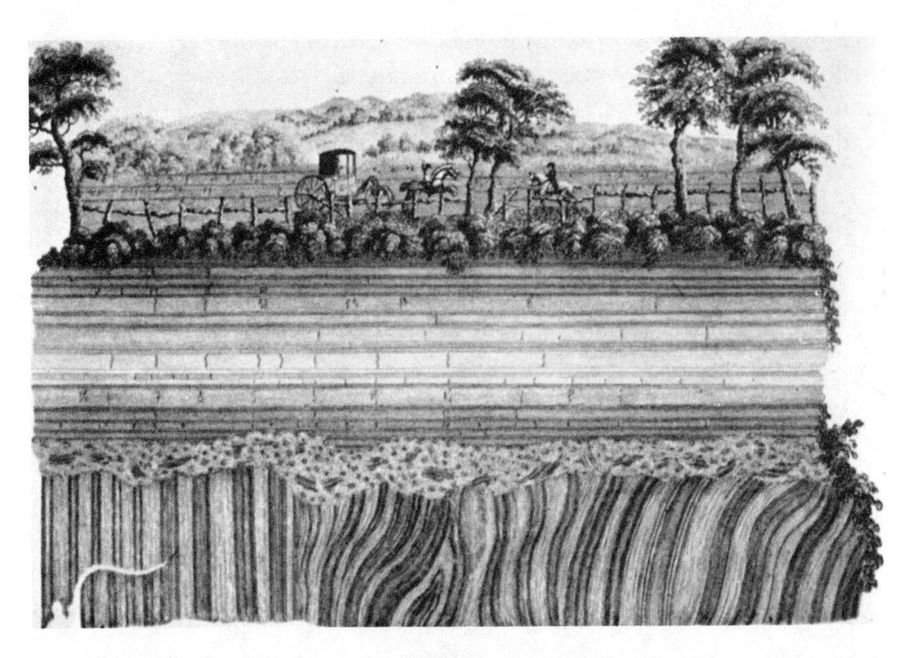

Unconformity at Jedburgh, Scotland, showing near-vertical beds of sandstone and shale (Silurian) overlain by flatlying beds of sandstone (Devonian). Between these two units is a layer of angular detritus composed of fragments derived from the older one. This is the same unconformity as the one exposed at Siccar Point. (Facsimile of engraving included in James Hutton's Theory of the Earth: the lost drawings, *published by the Scottish Academic Press, and reproduced here with the permission of Sir John Clerk of Penicuik, Bart.)*

Compare that flat account with what Playfair remembered about this same visit to Siccar Point. Commenting on the discordance between the red sandstone and the dark "schistus," Playfair recalls:

> On us who saw these phenomena for the first time, the impression made will not easily be forgotten. . . . What clearer evidence could we have had of the different formation of these rocks, and of the long interval which separated their formation, had we actually seen them emerging from the bosom of the deep? We felt ourselves necessarily carried back to the time when the schistus on which we stood was yet at the bottom of the sea, and when the sandstone before us was only beginning to be

> deposited. . . . An epocha still more remote presented itself, when even the most ancient of these rocks, instead of standing upright in vertical beds, lay in horizontal planes at the bottom of the sea, and was not yet disturbed by that immeasurable force which has burst asunder the solid pavement of the globe. Revolutions still more remote appeared in the distance of this extraordinary perspective. The mind seemed to grow giddy by looking so far into the abyss of time. . . . (*32*)

That view from the brink did not cause Hutton's mind to "recoil"; for as he tells us, he returned to Edinburgh "perfectly satisfied" (*33*).

CHAPTER NINE

Father William

Now this old cove as I've heard said
Had a very sharp nose to smell fish
And he very soon saw that every bed
Held its own particular shell fish.

ANDREW RAMSAY, 1856

In the late eighteenth and early nineteenth centuries a new principle for determining the geologic ages of fossiliferous strata emerged from field investigations in France and England. This is the principle of faunal succession, which in simplest form states that within sequences of fossiliferous beds different kinds of fossils succeed one another in a definite order.

Faunal succession derives its chronologic significance from the principle of superposition of strata. Both Steno and Hooke had cited evidence that in many situations the conditions of preservation and the positions of fossils embedded within a given stratum indicate that the organisms responsible for the fossils lived during the interval of time when the sediment forming the stratum was deposited. Hence superposition may determine the relative ages not only of strata but also of the fossils inside. Hooke moreoever recognized that ammonites and other fossils in British strata evidently have no living counterparts; and he even speculated that fossils might provide "criteria of chronology." The question of the chronologic potential of fossils was not, however, to be resolved by speculation, but by grubwork—by collecting and comparing fossils, stratum by stratum, formation by formation throughout thick sequences of rock units whose relative ages could first be determined by their order of superposition.

Giraud-Soulavie, abbot of Nîmes, was one such collector (*1*). In a report on the natural history of southern France, published in 1779, he classified a sequence of mostly marine strata according to whether the fossils they contained have or do not have representatives living in the present seas. The oldest strata, he announced, contain fossils such as ammonites and belemnites whose analogues do not live today; by contrast, the youngest strata contain species of shellfish, fishes, and plants living at present; while a sequence intermediate in position between these two contains a mixture of extinct and modern species. The picture of faunal succession that Soulavie painted with so broad a brush his fellow countrymen Georges Cuvier and Alexandre Brongniart depicted in fine detail in 1808. Their *Essay on the mineral geography of the Paris region* demonstrated that not only can whole formations be distinguished by their fossils, but in many instances individual strata themselves have distinctive faunal assemblages (*2*).

Soulavie was a learned clergyman, Brongniart a professor of natural history, Cuvier a prominent statesman and zoologist. But the person who deserves the greatest credit for demonstration of the principle of faunal succession was a man with the not unusual name of William Smith, self-educated son of a British blacksmith (*3*).

Smith was born in 1769 at Churchill, Oxfordshire, where he received elementary training in the village school. Early in his studies he displayed proficiency in drawing and geometry. These talents he put to use in teaching himself from textbooks the elements of surveying. Perhaps he knew that the services of surveyors were much in demand around Oxfordshire and adjacent counties. In any case Smith had good reason to acquire a marketable skill, for when he was seven years old his father had died, leaving him the eldest of five children.

At the time Smith was growing up in Churchill, England was also growing up industrially and agriculturally, and both of these developments created jobs for surveyors. The growth of industry was powered by coal, canals were required for transportation of this bulky fuel from the mines to the factories, and surveyors were needed to map the mines and lay out the courses of the canals. The "canal age" in England began around 1760 and culminated around 1815, by which year a network of some 2,200 miles of artificial waterways had been completed. At the same time, the major landholders were consolidating their scattered plots of pasture and farmland into contiguous fields that could be surrounded by hedges and ditches. Surveyors were needed to establish the metes and bounds of these consolidated properties.

In 1787 Edward Webb, surveyor at Stow-on-the-Wold, found that he had more business than he could handle, and so took Smith into his family as apprentice. Smith was eighteen years old when he entered this employment, and during the four years he worked for Webb he acquired the proficiency needed to establish himself as surveyor and civil engineer. In the course of travels about the countryside, which were a necessary and to him an enjoyable part of his training, he found opportunity to indulge a childhood hobby of collecting fossils.

When Smith was twenty-two he set up his own business in the coal-mining district of northern Somerset, with headquarters in a farmhouse at High Littleton a few miles west of Bath. His initial work there was in surveying the underground workings of the mines and in calculating their reserves. In the course of this work he became familiar with the sequence of coal-bearing strata and the younger non-productive beds superimposed upon them. His work must have been satisfactory, because in 1794, when the mine operators in Somerset began to entertain the idea of connecting their collieries by canal with Bath and thus to other parts of England, Smith was engaged to make a preliminary survey. Also he was commissioned to make a six-weeks tour to the north of England to learn more about the construction of canals. During this journey Smith recognized time and again the northward continuations of the strata he had come to know from his work around Bath.

For five years, beginning in 1794, Smith superintended the construction of the Somerset Coal Canal. In the course of excavations to the east of High Littleton, the canal crossed strata that lay above the coal-bearing sequence. Some of these younger formations were so similar in lithology that Smith had trouble in distinguishing them. On examining the fossils contained in this troublesome sequence, however, he discovered that each formation contained distinctive species. Isolated exposures could therefore be identified according to their proper position in the stratigraphic sequence by identification of the fossils they contain. This principle of faunal sequence he discovered sometime before 1796 (*4*). For some reason, Smith did not publicize his discovery immediately. Perhaps he was too busy with his work on the canal.

In 1796 Smith was elected to the exclusive membership of the

William Smith (Portrait courtesy of the Geological Society of London)

Bath and West of England Society. Through his associations with this group he developed a friendship with the Reverend Benjamin Richardson and through him with the Reverend Joseph Townsend. Both of these clergymen were interested in natural history, and both had collections of minerals and fossils. They were impressed to find that Smith could examine the fossils in their cabinets, and for each specimen collected in the area around Bath, tell them from which formation it had been taken. How Smith's knowledge of the sequence of formations and fossils was first reduced to writing is told by his nephew and pupil, John Phillips.

> One day, after dining together in the house of the Rev. Joseph Townsend (29 Pulteney Street, Bath) it was proposed by one of this triumvirate (Smith, Richardson and Townsend) that a tabular view of the main features of the subject, as it had been expounded by Mr. Smith, and verified and enriched by their joint labours, should be drawn up in writing. Richardson held the pen and wrote down from Smith's dictation, the different strata according to their order of succession in descending order, commencing with the chalk, and numbered, in continuous series, down to the coal, below which the strata were not sufficiently determined. . . .

> To this description of the strata was added, in the proper places, a list of the most remarkable fossils which had been gathered in the several layers of rock. The names of these fossils were principally supplied by Mr. Richardson. . . . Of the document thus jointly arranged, each person present took a copy, under no stipulation as to the use which should be made of it (later denied by Smith), and accordingly it was extensively distributed, and remained for a long period the type and authority for the descriptions and order of superposition of the strata near Bath. (5)

Smith's table lists twenty-three rock units in a vertical column. Parallel columns contain entries for the thicknesses, fossils, lithology, and other characteristic features of each unit. Most of the names Smith gave to his formations are informal indicators of the kind of rock present, such as Chalk, Sand, Clay, and Coal; but some names refer variously to localities (*e.g.* Pennant Street), physiographic features (*e.g.* Cliff), color (*e.g.* Red Ground), or utility (*e.g.* Fuller's Earth). For the unit below the Fuller's Earth, Smith could think of no apt description, so he christened it "Bastard ditto and Sundries." The names given to the fossils by Richardson were also informal. For example, the fossils listed for the Bastard Fuller's Earth and Sundries are "striated cardia, mytilites, anomiae, pundibs and duck-muscles" (*sic*). Nevertheless these informal names for formations and fossils stand for hard facts: a set of formations arranged in chronologic order according to the principle of superposition, and containing different kinds of fossils at different horizons. The time would come when paleontologists would give elegant Latinized names to the duck mussel and the prickly cockle (and then rename them about every twenty years) but the shapes and the stratigraphic positions of the fossils would not change thereby. A pundib by any other name. . . .

Smith dictated this stratigraphic table shortly after he lost his job with the Somerset Coal Canal Company in June of 1799, evidently as a result of a squabble with the managers over business affairs. Relieved of the routine of daily work, Smith found time to complete a map showing the areal distribution of geological formations within a radius of five miles around the city of Bath. This has been called the oldest geological map, as distinguished from earlier soil maps (*6*). Smith was here interested primarily in bedrock, not in soils and other surficial materials that mantle much of England. Using the information provided by exposures of bedrock in natural croppings, in mines and wells, and in excavations along canals, he was able to infer the traces of the *contacts* between the formations, and thus to show in different colors the areas underlain by different formations, as though the soil and other surficial materials had been swept away.

Smith's work with the Coal Canal Company had gained him a local reputation as civil engineer sufficient to insure his livelihood as a consultant. One of his early assignments was to arrest landslides which in 1799 threatened to destroy buildings in Bath. Smith's approach to the problem was to drive tunnels into the base of the unstable slopes, thus draining out the groundwater that provided a lubricant for the slides. Thereafter he engaged in a wide variety of projects including the draining of swamps and the irrigation of farmlands, surveying coal mines, prospecting for coal and groundwater through borings, and devising plans for the construction of canals. In 1802 he moved his headquarters to Bath, and two years later he established his residence in London, in a four-storeyed house renting for 80 guineas per year.

By 1810 Smith's reputation as engineer was more than provincial. In 1806 he published his first book, a treatise on agricultural engineering related to the management of marshy meadows and to the draining of peat bogs (7). As announced in the lengthy title, this work gives "an account of Prisley Bog and other extraordinary improvements conducted for His Grace, the Duke of Bedford; Thomas William Coke, Esq. M. P. and others; by William Smith, Engineer and Mineralogist." Modesty was not one of Smith's faults.

In 1809 the hot mineral springs for which Bath has been famous since the days when the Romans dedicated them to the goddess Minerva began to fail. Smith was called upon to investigate. He quickly determined that the hot waters were still flowing underground, but were by-passing the old vent through new channels. Those passages he caused to be plugged, whereupon the waters flowed back into the Pump Room and Baths as before. For this accomplishment Smith received a handsome stipend and much favorable publicity.

From his headquarters in London, Smith travelled to various points in England and Wales to fulfill his commissions as consultant. Often he proceeded to his destinations not directly but by circuitous routes, in order to expand his network of geological traverses. Thanks to the fact that fast mail coaches were in service after 1784, he could log as many as 10,000 miles in a single year. The records clearly show that Smith did all this bumping about in zigzag courses with one compelling purpose in mind: to publish a geological map of all England and Wales. As early as 1801, about a year before he moved to Bath, he had prepared a prospectus for a work entitled *Accurate delineations and descriptions of the natural order of the various strata that are found in different parts of England and Wales: with practical observations thereon.* The text promises that the single quarto volume will be illustrated by "a correct map of the strata," each formation to be rendered in a distinctive color (*8*). Subscribers could have the book for two guineas, payable on delivery. The startling thing about all this is that delivery was promised only five months after the prospectus was issued. More surprising still, Smith at age thirty-two already had in hand a mansucript map with the areal geology of most of England and Wales colored in. He must have calculated that he could map the geology of East Anglia, the London Basin, and the Yorkshire moors, and in his spare time write the text, all in five months' time. The prospectus was printed, but the book was never completed, and the project came to an end when Smith's publisher declared bankruptcy.

Fourteen years would pass before Smith's geological map was published. The final product, entitled *A delineation of the strata of England and Wales with part of Scotland*, consisted of fifteen sheets, scale five miles to the inch (*9*). When fitted together the sheets form a map six feet across and eight and a half feet tall. Twenty-three formations are distinguished, and the areas of their outcrop are hand-colored. The first sheets were exhibited in 1814, and upwards of 400 copies of the entire map were issued in 1815 and afterward. In addition to the geology, the map shows the location of canals, roads, railroads, and mines. Smith's map was the first to represent on a large scale the geological relations of so extensive an area; and it set a style for depicting areal geology that has persisted with few changes until today.

In 1815 Smith also published a memoir describing the formations shown on his map and the kinds of soils developed on the various units. The introductory passages contain his first published statement concerning the orderly arrangement of strata and fossils which he had discovered.

> . . . I presume to think that the accurate surveys and examinations of the strata, . . . near the surface of the earth and in its interior, to the

> greatest depths . . . penetrated by the sinking of wells, mines, and other excavations, to which I have devoted the whole period of my life, have enabled me to prove that there is a great degree of regularity in the position and thickness of all these strata. . . . ; and that each stratum is also possessed of properties peculiar to itself, has the same exterior qualities, and the same extraneous or organized fossils throughout its course. (*10*)

Note that in the above passage Smith's "stratum" corresponds in modern use to "formation" as does his "strata" to our "formations." "Extraneous fossils" and "organized fossils" are now simply referred to as "fossils"; in Smith's day they were also called "petrifactions."

To substantiate his claim that the different rock units shown on his map contain different kinds of fossils, Smith began publishing a work that would show pictures of the fossils characteristic of each formation. The first of a projected seven-part series, entitled *Strata identified by organized fossils*, appeared in 1816, and the fourth and last part to be issued is dated 1819 (*11*). The fossils are illustrated by carefully executed drawings on nineteen plates, each printed on paper tinted to match the approximate color of the formation from which the specimens had been collected.

The reason why this picture book was never completed was that Smith was bankrupt in 1819. His financial troubles began in 1811 when a business venture in the quarrying of building stone failed and left him in debt. Although his consulting fees were sufficient to insure a good living, he invested heavily in publishing his map and other works, the sales of which were not substantially rewarding. In 1815 he began arrangements for selling his collection of some 2,600 specimens of fossils to the British Museum, and two years later he published a description of that collection under the title *Stratigraphical system of organized fossils, with reference to the specimens of the original geological collection in the British Museum* (*12*). The sale of the fossils had brought £700, but this combined with consulting fees and income from sales of maps and books was still not enough to satisfy his creditors. Finally in June of 1819 his financial losses led to loss of his freedom as well, and he was sent to Kings Bench Prison (*13*). After more than two months there he obtained his release, sold his furniture and books, and moved to the north of England.

For the next ten years, Smith moved about in Yorkshire and adjacent counties, lecturing, surveying, and publishing maps, articles, and broadsheets. Records remain of his geological lectures before philosophical and literary societies at Hull, Sheffield, Leeds, Scarborough, and elsewhere. For a time he served as Resident Land Steward to a gentleman farmer of Hackness. Prior to departing London, he had begun yet another grandiose project, a "New Geological Atlas of England and Wales." The first part, covering four counties, had been published prior to his imprisonment; and the second, covering four more, appeared a month after his release. Another of Smith's unfinished products, the last part appeared in 1824, bringing the number of counties covered to twenty-one.

Smith's scientific work was accomplished during the interval of time when geology emerged as an independent science in England. In 1807 the Geological Society of London was organized. This earliest of geological societies was founded by a group of gentlemen whose ties of friendship were reinforced by a common interest in the history of the earth, and who had the general education, means, leisure, and brains needed to pursue this avocation (*14*).

Smith was not a member of the Geological Society. He was one in a number of mostly self-educated individuals who earned their living and respect by superintending the construction of canals, the development of coal mines, the draining and irrigation of land, and a host of other undertakings of an economic nature.

The time came, however, when members of the Geological Society felt obliged to acknowledge their debt to Smith. In 1831 the Society presented him with the Wollaston Medal, the first award of this its highest honor. In his presentation address the president, Adam Sedgwick, mentioned how much he personally owed to Smith.

> I for one can speak with gratitude of the practical lessons I have received from Mr. Smith; it was by tracking his footsteps, with his maps in my hand, through Wiltshire and the neighboring counties, where he had trodden nearly thirty years before, that I first learnt the subdivisions of our oolitic series, and apprehended the meaning of those arbitrary and somewhat uncouth terms, which we derive from him as our master, and which have long become engrafted into the conventional language of English geologists. . . . (*15*)

Sedgwick went on to proclaim Smith the *Father of English Geology*. As justification, he observed:

> If we, by our united efforts, are chiseling the ornaments and slowly raising up the pinnacles of one of the temples of Nature, it was he that gave the plan, and laid the foundations, and erected a portion of the solid walls by the unassisted labour of his hands. . . .

This was to be only the first rechristening. Other distinguished geologists, looking back on Smith's career in longer perspective of time, have called him the *Father of Stratigraphy* and even the *Father of Historical Geology* (*16*). Father William, indeed! And *Dr.* Smith at last, after Trinity College, Dublin, conferred upon him the LL.D. in 1835.

Now in failing health Smith retired to a cottage at Scarborough, where he was sustained in part by his yearly pension of £100 from the Crown, first granted in 1832. In 1838 he was appointed to a commission to select stone for rebuilding the Houses of Parliament, which had that year been destroyed by fire. In 1839 he examined and appraised a collection of fossils in the custody of the Literary and Philosophical Society of Hull. A few months later he died at Northampton at age seventy while en route to Birmingham for a meeting of the British Association for the Advancement of Science. Contrary to some accounts, Smith did not die unmarried: he had a wife, but no one seems to know who she was or when he married her (*17*).

Contemporary accounts of Smith do not tell much about him as a person, beyond the impressions that he had the face of a farmer and the hale physique to match. We are told that he was bluff in manner, restless, energetic, always ready to talk about his ideas to anyone who would listen. He seemed always under pressure to meet his business commitments, and was often pressed for lack of funds. In his scientific work he was a "loner," in the sense that he would direct his own investigations, however much he might discuss them with others.

Smith's science was empirical, at least on the surface, and practical too. Time and again he emphasizes the utility of geologic investigations. In his memoir of 1815, he explains that his map will save the useless expense of prospecting for natural resources where they are not to be found. A better knowledge of geology, he insists, will lead to improvements in mining, in agriculture, and in the laying out of canals, roads, and railways. Miners will be guided to places that will yield metals and coal; builders to freestone and brick-earth; inhabitants of dry places to water; and the farmer to mineral fertilizers.

Utility aside, the pursuit of geology is an entertaining and healthful exercise, Smith insisted. "Rural amusements," he wrote, "to those who can enjoy them, are the most healthful; and the search for a fossil may be considered at least as rational as the pursuit of a hare" (*18*). And these amusements are good for young and old alike. Fossil collecting is "well calculated for the healthful and rational amusement of youth" and "Natural History should be the first object of every country gentleman."

Beyond utility and health-giving entertainment, the regular chronologic sequence of fossils in strata discloses yet another example of order in nature, according to Smith.

> Fossils have been long studied as great curiosities, collected with great pains, treasured with great care and at a great expense, and shown and admired with as much pleasure as a child's rattle or a hobby horse is shown and admired by himself and his play fellows, because it is pretty; and this has been done by thousands who have never paid the least regard to that wonderful order and regularity with which Nature has disposed of these singular productions and assigned to each class its peculiar stratum. (*19*)

In the preface to his *Stratigraphical system of organized fossils* (1817), Smith wrote: "My observations on this and other branches of the subject are entirely original, and unencumbered with theories, for I have none to support" (*20*). The reference here is obviously to theories of a scientific nature; and indeed there is nothing in his writing to suggest that he was guided to his discovery of faunal succession by any idea of organic evolution. On the other hand there is much to indicate that he interpreted his science according to certain metaphysical theories which were commonly held in his time.

Smith was a creationist: he held that the different kinds of organisms whose remains are found at different levels in the strata were supernaturally created and supernaturally extinguished at successive intervals of time. He was also a progressionist; his successive creations were not random but progressive in quality of life. As he put it, "In this as in every other part of creation, there seems to have been one grand line of succession, a wonderful series of organization successively proceeding in the same train toward perfection" (*21*).

With regard to successive creations, Smith was most explicit in a broadsheet he published in 1835 entitled *Deductions from established facts in geology* (*22*). Here he recognized nine stages in the earth's development. During the first period, when the oldest rocks formed, there was no life on earth. During the second, third, and fourth stages organisms of different kinds successively appeared and disappeared in an alternating series of supernatural creations and extinctions. Land first emerged from the sea during the fifth stage. Stage 6 is marked by "THE CREATION," presumably of the antediluvian plants and animals, and stage 7 by "the great supernatural event of the Deluge," whose effects are known from surficial accumulations of water-worn stones containing bones. Stage 8 was marked by an episode of refrigeration, as evidenced by "an entire elephant preserved in the ice of Siberia." We are living in Stage 9.

Smith was a providentialist in that he believed the earth was created for the habitation of man. Everything about the strata is useful. The orderly succession of fossilized organisms in the strata "must strike the admirers of nature with a degree of reverential awe and grateful adoration of the Almighty Creator," the one and only destroying and animating power. On the practical side, this succession permits strata to be identified according to the fossils they contain, thus providing guides in the search for natural resources hidden underground. The very fact that the strata are not arranged in concentric, onion-skin fashion but rather are inclined in different directions so that their edges are exposed at the surface represents "one of the most wonderful dispositions of Providence that could be designed for the benefit of man" (*23*). For otherwise coal and mineral deposits could not be mined, nor could any but the youngest strata be identified by fossils.

In his writings Smith does not speculate on the age of the earth. He was, however, impressed with the cumulative bulk of all the fossils embedded in the stratigraphic column, which led him to observe that "the consideration of such an immensity of animal and vegetable matter, the time required for its perfection and subsequent consolidation in the strata, evidently in deep and quiet waters, may seem at first incomprehensible to many." But the passage ends lamely with words to the effect that anything is possible under the guidance of the Creator (*24*).

However evasive Smith might be regarding the age of the earth, his synopsis of geologic history in the broadsheet of 1835 implies that the successive events of creation and destruction required a long stretch of time that might indeed be incomprehensible. THE CREATION comes late in his chronicle; and the Deluge is no more than an episode in earth history, a supernatural event in a long chain of such events.

The discovery and documentation of faunal sequence by Smith and his contemporaries provided the means of establishing a standard column of strata, first for England and Western Europe and later for all the world. Not all parts of this column would be found at any single locality: local columns at a host of localities would be correlated and pieced together, part by part, on the bases of their order of superposition and the similarities in their fossils. After the order of succession of the fossils had become established, fossils could be used as independent evidence for the geologic age of strata.

Despite all metaphysical preoccupations, Smith remained in matters of science an empiricist, a Baconian. Witness his verse.

> Theories that have the earth eroded
> May all with safety be exploded
> For of the Deluge we have data
> Shells in plenty mark the strata,
> And though we know not yet awhile
> What made them range, what made them pile,
> Yet this one thing full well we know—
> How to find them ordered so. (*25*)
>
> (SMITH, 1829)

CHAPTER TEN

The Four-dimensional Jigsaw Puzzle

Order is Heaven's first law.
ALEXANDER POPE

Long before faunal succession was recognized as a chronologic principle, European geologists had attempted to divide the rocks of the visible crust into a few major units differentiated mainly on composition, and ordered in age according to their order of superposition. Each of these subdivisions contained many smaller rock units of the kind Smith delineated on his great map of 1815.

In 1756 Johann Gottlob Lehmann (1719–1767) proposed a threefold classification of rocks, based upon his studies in Thuringia and other parts of northern Europe (*1*). Though Lehmann was trained as a physician, his interests inclined more toward science and engineering. He taught courses in mining and metallurgy in Berlin; and later, by appointment of the Empress Catherine, he served at St. Petersburg as professor of chemistry and director of the Imperial Museum.

In the Harz Mountains, Erzgebirge, and surrounding hilly areas, Lehmann perceived a connection between the prominence of the mountains and the kind of rocks forming them. The central parts of the two mountainous belts expose granite and other crystalline rocks, which Lehmann designated as *Primitive.* Around the flanks of the mountains these primitive rocks are overlapped by flat-lying or gently tilted strata, which Lehmann called the *Floetzgebirge* (a miner's term signifying horizontally stratified rocks). The second group, in turn, is locally covered by unconsolidated gravel, sand, and silt, the *Alluvium.*

Lehmann sought to account for these three classes of rocks in terms of a neptunian theory, which resembles de Maillet's theory in certain respects. In the beginning, he conjectured, the materials forming the Primitive Rocks were suspended as sediment in the turbulent waters of a universal ocean. After the turbulence had decreased, sediment settled out unevenly on the sea floor; and as the waters retired into the basins of the oceans and seas, the higher heaps of sediments emerged as the primitive mountains.

Noah's Flood was the second great event in Lehmann's chronicle. The floodwaters overtopped the primitive mountains, and on retreating swept the earth from these and redeposited it as the Floetz strata. Remains of terrestrial and aquatic organisms which had lived on the slopes of the primitive mountains, or in lakes, were swept out to sea by the subsiding floodwaters and incorporated in the Floetz.

The third and youngest class of rocks Lehmann considered to be the least important. Some of these deposits, formed from reworked Floetz materials, he thought might record the final effects of the Deluge or of more recent floods.

At the same time that Lehmann was conducting his investigations in northern Europe, Giovanni Arduino (1714–1795) was working on a classification of mountains based on the relative ages of the rocks that form them. He conducted his field studies in the Alpine foothills around Padua, Vicenza, and Verona, and in the high mountains to the north. Like Lehmann, Arduino was interested in mining: he served as an official of the governmental mining service before accepting appointment in Venice as professor of mineralogy and metallurgy. His principal geological findings were published in 1759 and 1760.

According to Arduino, the rocks forming the earth's crust consist of four major units. The oldest are the mineral-bearing crystalline rocks that form the *primary* or *primitive mountains.* Overlying these, and hence younger, are the rocks of the *secondary mountains:* sequences of marble and stratified limestone which are mostly lacking in mineral deposits but which contain abundant marine fossils. The *tertiary mountains* are still younger in age and lower in relief; they are formed of gravel, fossiliferous sand and clay, and volcanic materials. The fourth and youngest unit consists of sediment washed down from the mountains by streams, that is to say of *alluvium.*

The classifications proposed by Lehmann and Arduino are significant on several counts. Both investigators recognized that the rocks

Abraham Gottlob Werner

forming the earth's crust may be divided into a few major units on the basis of composition, presence or absence of stratification, and presence or absence of fossils. These major units form a chronologic sequence, as established by their order of superposition. Since mountains can be no older than the materials that form them, and since different ranges of mountains are formed of different units, it must follow that some mountain ranges are older than others. All the mountains in the world cannot then be explained as the result of a single event, such as the catastrophic collapse of the crust as envisioned by Burnet (*2*). Arduino's term, *Tertiary*, would later take its place in the standard stratigraphic column, now recognized by geologists the world over. Lehmann's classification formed a basis for the work of the greatest neptunist of them all, Abraham Gottlob Werner (*3*).

Werner was born in 1749 at Wehrau, Saxony (near Gorlitz on the present border between East Germany and Poland). His father, who managed the iron works of the Duke of Solm, gave the boy early instruction in mineralogy. When Werner was fifteen, his mother died, and he dropped out of school to work for his father. At the iron works he did a variety of chores, including the assaying of ores and keeping the company's books of account. After five years of this work he went to Freiberg, where he studied mining and metallurgy at the Bergakademie, a school of mining established in 1765. In 1771 he entered the University of Leipzig to study law, but his interests soon changed after he had elected courses in a variety of other subjects including natural science, history, and languages. Also he pursued independent studies of mineralogy, and while still at Leipzig wrote a book on the external characters of minerals which was published in 1774 (*4*).

The book on minerals brought instant fame to Werner when he was only twenty-five. During the year of its publication, he accepted an appointment at the school of mining in Freiberg; and he remained at this post until his death in 1817—a tenure of forty-two years.

Werner's success as a teacher can only be described as phenomenal. His courses in mining, mineralogy, and related subjects attracted bright students from distant parts of Europe. Several of the approximately thousand persons who studied with Werner later made significant contributions to geology. Among these stars Alexander von Humboldt is perhaps the most highly regarded by modern scientists.

Werner's theory of the earth is known only in part from his own publications. His *Short Classification and Description of the Different Rocks,* which appeared in 1786, recognizes four categories (*5*). The oldest, or *Primitive Rocks,* consist of crystalline substances of the kinds called granite, porphyry, basalt, gneiss, serpentine, marble, or quartzite. Fossils are lacking, but some primitive rocks, such as granite, contain ores of tin and iron. The *Floetz Rocks* overlie the Primitive, and include stratified units of limestone, sandstone, clay, chalk, coal, rock-salt, and gypsum. They differ from the Primitive Rocks in that they contain fossils, and are more limy and clayey. Within a single formation of the Floetz, several different kinds of rocks may alternate with each other, though one type always predominates.

Werner's third category, *Volcanic Rocks,* he divided into two parts. The "true volcanic rocks" include surficial deposits of lava, pumice, ash, and the solidified ash called tuff. Commonly these materials form conical mountains with craters at the top. "Pseudo-volcanic rocks" form lava-like layers in the Floetz sequence, but actually these have formed by the combustion of coal beds, so Werner insisted.

The *Alluvial Rocks*, youngest of the four types, consist almost entirely of materials reworked from the Primitive or Floetz, and sometimes even from the Volcanic Rocks. In mountainous regions these deposits consist mainly of gravel and sand flooring the valleys. Sand, loam, bog deposits, and peat belonging to this same unit are commonly spread over the surfaces of lowland plains. The Alluvium locally contains petrified wood and bones of quadrupeds, but many of the fossils in it have been reworked from the Floetz.

With increase of information, Werner revised his lectures, though not his basic ideas. About ten years after his book on rocks was published, he introduced to his system a new unit which he called *Transitional*, inserted between the Primitive and the Floetz. In Werner's view, this unit resembled the Primitive in that it consisted mostly of chemical precipitates from water; on the other hand, it resembled the Floetz in that it contained mechanical or clastic sedimentary deposits and abundant fossils.

To explain this fivefold sequence, Werner assumed that in the early stages of the earth's development an uneven and solid nucleus was submerged beneath a universal ocean. The waters of this ocean contained all the materials that form the present crust. Some of these materials were dissolved in the water, others were suspended as solid particles. At a stage when the waters of the ocean were calm, crystals of various minerals were precipitated and settled on the bottom to form the Primitive Rocks. First a blanket of granite was deposited, its upper surface reflecting the depressions and elevations on the nucleus below. Other kinds of crystalline rock were laid down in turn, some of these intermingling with the granite. Concurrently the water of the ocean was diminishing in volume, so that the level of the sea was lowered. Werner conjectured that this lowering was due to decomposition of water which enriched the atmosphere or perhaps escaped into outer space.

With the continued lowering of the sea, solid particles suspended in the water began to settle along with the chemical precipitates to form the Transitional Rocks. At some point during the Primitive or Transitional periods islands emerged from the receding ocean, but at first these landmasses were small and so produced minor discontinuities in the blankets of older sediment. Thus Werner considered all the Primitive and much of the Transitional sequences to be essentially universal in extent.

As the seas continued to shrink, the predominantly clastic and richly fossiliferous Floetz strata were deposited, and after them the even more geographically restricted Alluvial Rocks. Combustion of coal beds led to sporadic and localized volcanic activity.

No brief sketch such as this does justice to the flexibility of Werner's theory, nor to its plausibility in the minds of many of his contemporaries and most of his students (*6*). Werner did not insist that his major categories of rocks are all mutually exclusive, but freely admitted that all (with the exception of the Volcanic) are intergradational. Thus some Primitive rocks could almost as well be classified with the Transitional, and some Floetz rocks with Alluvium. Nor did he hold that the diminution of the ocean was a constant and uninterrupted process; rather, he allowed for fluctuations in level and variations in turbulence. These fluctuations he called upon to account for overlap of certain younger formations above older. In explaining the steep dips and folds commonly observed in stratified rocks, he appealed to the kinds of processes that operate today. Chemical precipitates need not obey the principle of original horizontality; they may accumulate on steeply inclined surfaces. Thus a steeply dipping formation of marble, conceived as a mass of chemically precipitated crystals, might simply have been plastered onto a steeply inclined substrate and therefore not have been bent from an original horizontal position. The deformation of clastic rocks, disposed in folds or broken by faults, might be explained as a result of submarine slumping when the beds were still soft.

With regard to the antiquity of the earth, Werner was vague; but there is nothing in his writings to suggest that he subscribed to the Biblical chronology. In a popular lecture he delivered in 1817 he observed that "our earth is a child of time and has been built up gradually" (*7*). Referring to the four categories of rocks described in his *Short Classification*, he accounts for the gradations between them as expectable in view of "the vast period of time since the beginning of our earth" (*8*). Again, in an unpublished manuscript, he writes of the "enormously great time spans, which perhaps far exceed our imagination of the time of development of the rock masses" (*9*).

The Lehmann-Werner classification of major rock units was widely accepted in Europe during the late eighteenth and early nineteenth centuries. This scheme was also applied by William McClure

(1767–1840) to the classification of rocks in the United States (*10*). McClure, who has been called the father of American geology, published geological maps of the eastern United States beginning in 1809. In true Wernerian fashion the crystalline rocks are classified as Primitive, the steeply dipping stratified rocks as Transition, the essentially horizontal strata as Floetz or Secondary, and the unconsolidated or weakly consolidated sediments of the lowland areas as Alluvial.

Around the beginning of the nineteenth century, European geologists became engaged in a systematic classification of fossiliferous strata into units of magnitudes intermediate between formations and the vast sequences of formations encompassed in the Transitional and Floetz units. These new units, or *systems*, were distinguished mainly on the basis of the kinds of rocks that formed them and were ordered chronologically according to superposition (*11*).

Thus in 1799 Alexander von Humboldt applied the name *Jurassic* to the predominantly limestone sequences exposed in the Jura Mountains of Switzerland and France. To the west of these mountains the hard Jurassic limestones dip beneath sequences characterized by deposits of chalky limestone, for which the Belgian geologist Omalius d'Halloy proposed the name *Cretaceous* in 1822. In and around Paris the Cretaceous strata are overlain by interbedded sand, clay, marl, limestone, and gypsum to which Arduino's name *Tertiary* was applied. Mindful of the Arduino scheme, the French geologist Jules Desnoyers in 1829 applied the name *Quaternary* to deposits overlying the Tertiary in the Paris Basin and previously designated as "Alluvium."

North of the Jura Mountains the strata underlying the Jurassic system are exposed over wide areas on either side of the Rhine Valley. In Germany these underlying strata consist of limestone sandwiched between reddish-brown sandstone and clay. Friedrich von Alberti's designation of these strata as *Triassic* in 1834 reflects the threefold nature of this unit.

Of the six systems that are older than the Triassic, five were named for localities in England and Wales. In 1822, two British geologists, William Conybeare and William Phillips, proposed the name *Carboniferous* for the coal beds and associated strata of north-central England. The strata so designated fall naturally into two parts: a younger coal-bearing sequence (Upper Carboniferous), and an older sequence composed variously of sandstone and limestone (Lower Carboniferous).

Adam Sedgwick (Portrait courtesy of Harry B. Whittington, Sedgwick Museum, University of Cambridge)

The succession of the older Transition beds of Werner, as exposed in Wales and along the Welch border, had eluded William Smith. Here in this rugged terrain the strata are bent into folds and broken by faults, and wherever fossils are present they normally do not weather from the hard rock but must be hammered or chiseled from it. In the early 1830's the Reverend Adam Sedgwick and Sir Roderick Murchison set out to put in order the stratigraphy of this region. Sedgwick began his work in the northern part of Wales, Murchison in the south. After three summers of fieldwork, Murchison had been able to distinguish and place in chronologic order according to their superposition a sequence of rock units below the base of the Carboniferous System of Conybeare and Phillips. Moreover he could demonstrate that each of these units contained distinctive fossils. In 1835 he named these collective strata the *Silurian System*, after the Silures, a native tribe that fought a losing contest with the Romans beginning in A.D. 48 (*12*).

Meanwhile Sedgwick had established a tripartite sequence of rock units, including slate, volcanic rock, and mudstone. In a joint paper with Murchison, also published in 1835, he named this sequence the *Cambrian System*, after Cambria, the Latin name for Wales (*13*). Unlike Murchison, he had neglected to study the fossils in the Cambrian. At the time of the joint presentation, however, both investigators were satisfied that the upper part of Sedgwick's Cambrian and Murchison's Lower Silurian were contiguous and hence sequential in age. This opinion soon had to be revised as further studies showed that the two systems in fact overlapped. Thereafter a contest developed between Sedgwick and Murchison to determine whose system should be robbed of the strata held in common under different names. Neither man was inclined to yield. The issue was finally settled, after the deaths of the contestants, by an industrious Scot. Charles Lapworth marshalled evidence to show the Cambrian-Silurian sequence actually contained three distinct assemblages of fossils; one in Sedgwick's Cambrian, a second in Murchison's Lower Silurian, and the third in the Upper Silurian. Accordingly he removed the Lower Silurian from the previous classifications and

Roderick Impey Murchison (*Portrait courtesy of the Geological Society of London*)

rechristened it the *Ordovician System* (*14*). The name comes from the Ordovices, a British tribe that once lived in North Wales.

In 1839 when Sedgwick and Murchison were still on friendly terms, they jointly designated as *Devonian* (from exposures in Devonshire) the strata between the Silurian and Carboniferous Systems (*15*). In so doing they appropriated to their new system the Old Red Sandstone, a prominent formation which Conybeare and Phillips had earlier included in the Carboniferous.

In 1840 and 1841 Murchison extended his stratigraphic studies into eastern Europe. Travelling across Russia and down the western side of the Ural Mountains he found sequences of strata containing the same successions of fossils as had been established for the Silurian, Devonian, and Carboniferous systems of England. Above the beds containing typical Carboniferous fossils were thick accumulations of marl, limestone, sandstone, and conglomerate containing a flora and fauna sufficiently distinctive to set this sequence apart both from the Carboniferous below and the Triassic above. Accordingly, Murchison named these strata the *Permian System*, after the town of Perm on the western edge of the Ural Mountains. (*16*).

The work of classifying the abundantly fossiliferous rocks of the earth's crust according to geological systems required one hundred twenty years for its consummation, beginning with the designation of the Tertiary by Arduino and ending with Lapworth's establishment of the Ordovician. From any point of view the end result was a patchwork. With regard to the names given to the systems, Tertiary and Quaternary refer to chronologic order in a numerical scheme; Triassic to the number of rock units within a system; Cretaceous and Carboniferous to characteristic types of earth substances. The rest of the names reflect geographic location—either physically, as in the case of the Jurassic; politically, as for the Cambrian, Devonian, and Permian; or ethnologically, as in the Silurian and Ordovician.

More substantial differences pertain to what these names were originally intended to designate. At one extreme are names that refer to rocks: Humboldt's Jurassic and Sedgwick's Cambrian are examples. At the other are names based on distinctive assemblages of fossils in designated sequences of strata. Lapworth's Ordovician is a prime example, and his designation of this system necessarily entailed as well a redefinition of the top of the Cambrian and the base of the Silurian. Murchison's Permian System was established on the distinctiveness of its fossils as compared with those characteristic of the Carboniferous and Triassic sequences below and above.

By the time Murchison began his Russian excursion in 1840, geologists had come to realize that systems are not universal in geographic extent and are highly variable from region to region according to thickness and the kinds of rocks that form them. On the other hand, the fossilized plants and animals within the systems are sufficiently distinctive to form a reliable base for piecing together the disunited sequences of strata in different parts of the world.

More often than not, it has turned out that the localities at which the systems were first described were not the best for reference in reconstructing geologic history. For example, the type section for the Cambrian system is in northern Wales, where the strata are not only sparingly fossiliferous but also complexly deformed. A sequence of strata in southern Scandinavia which is relatively undisturbed by folds and faults, and which not only contains the same kinds of fossils found in Wales but many more besides, has been accepted as a reference section for this system.

Whether in or out of a type locality, however, systems are made of rocks, and the intervals of time required for these rocks to form have long been known as *periods*. Ideally, the periods corresponding to successive systems should be successive too, without gap or overlap. Obviously this was not the case when Sedgwick and Murchison named the Cambrian and Silurian Systems. Lapworth's establishment of the Ordovician System was designed to eliminate the overlap in strata, and hence by inference the overlap in time. Most adjustments in the original boundaries of the systems have been in the other direction—that is, to fill the gaps. Humboldt's section in the Jura Mountains, for example, now represents only a part of what is called the Jurassic System, as a result of adding strata from reference areas farther to the west. This process of amending boundaries continues with increase of information. In some instances a given rock unit cannot be assigned to some system with any certainty for lack of evidence. This fact is less discomforting when one considers that systemic classification is pretty much a matter of convenience in organizing information. Changing the boundary of a system will not produce an earthquake, though it might jeopardize a friendship, as Sedgwick and Murchison have proved.

Before the middle of the nineteenth century enough was known about the succession of fossils in the systems to show that this sequence is not random. Thus the Cambrian strata are characterized by an abundance of trilobites and brachiopods. Bryozoa and corals appear in the Ordovician. The Silurian contains the first authenticated fishes and land plants. The oldest amphibians are found in strata of Devonian age. Remains of highstanding trees, with woody trunks, first appear in the lower parts of the Carboniferous; and the oldest reptilian fossils are found in the upper parts of that system. Ammonites, the "serpent-stones" of the early naturalists, are characteristic invertebrate forms in the Triassic System, which also contains the oldest remains of mammals and dinosaurs. The Jurassic marks the appearance of modern bony fish and birds. Remains of flowering plants first appear in the Cretaceous sequence, which also contains the remains of the largest flesh-eating dinosaurs. The oldest known placental mammals occur in the Tertiary and the oldest remains of modern man in the Quaternary.

This hit-and-miss recitation of organic remains characteristic of the geological systems does scant justice to the record of life as preserved in the rocks, but it illustrates important differences that led to a clustering of systems into even larger units according to the kinds of fossils they contain. Based on proposals in 1838 by Sedgwick and in 1840 by John Phillips, William Smith's nephew, the systems Cambrian through Permian have been called the Paleozoic, as the repository of "ancient life." Mesozoic is a name for "medieval life" as preserved by fossils in the Triassic, Jurassic, and Cretaceous Systems. The Tertiary and Quaternary Systems constitute the Cenozoic or "modern life" segment of the stratigraphic columns.

Here again a distinction needs to be drawn between substance and time. Just as the material systems are the basis for the inferential *periods*, so are the Paleozoic, Mesozoic, and Cenozoic supersystems (terrains or erathems in modern terminology) the basis for *eras*. From the standard stratigraphic column of systems the following scale of geologic time is derived.

The geological time scale, as originally developed, was uncalibrated in terms of years. The units in the scale relate to relative ages only—to "older than" or "younger than" relationships between the strata that constitute the systems and terrains from which the periods and eras are inferred. These older-younger relationships ultimately depend on the principle of superposition for their validity, just as the principle of faunal sequence derives from order of superposition. This fact has been persistently ignored by individuals and groups who seek to discredit natural science on grounds of religion. For example, it is claimed that the standard stratigraphic column was put together on the *a priori* assumption of organic evolution (*17*). This is simply not true. Many geologists of the early nineteenth cen-

STANDARD STRATIGRAPHIC COLUMN *Stratigraphic Units*	GEOLOGICAL TIME SCALE *Time Units*
CENOZOIC TERRAIN*	CENOZOIC ERA
Quaternary System	Quaternary Period
Tertiary System	Tertiary Period
MESOZOIC TERRAIN	MESOZOIC ERA
Cretaceous System	Cretaceous Period
Jurassic System	Jurassic Period
Triassic System	Triassic Period
PALEOZOIC TERRAIN	PALEOZOIC ERA
Permian System	Permian Period
Carboniferous System**	Carboniferous Period
Devonian System	Devonian Period
Silurian System	Silurian Period
Ordovician System	Ordovician Period
Cambrian System***	Cambrian Period

*Terrain is the same as supersystem or erathem of some authors.

**In North America the Lower Carboniferous is usually called the Mississippian System, and the Upper Carboniferous the Pennsylvanian System.

***Efforts to establish systems older than the Cambrian have to date been unsuccessful because of scarcity of fossils in the oldest rocks forming the earth's crust. For these ancient strata the term Pre-Cambrian is commonly used.

tury held to a directional view of earth history, as opposed to a cyclical or steady-state view. But the Reverend Adam Sedgwick, writing in 1845, made the point abundantly clear that direction of changes in the organic world does not imply evolution.

> Now I allow (as all geologists must do) a kind of progressive development. For example, the first fish are below the reptiles; and the first reptiles older than man. I say we have successive conditions (so far proving design), and not derived in natural succession in the ordinary way of generation. (*8*)

Obviously, Sedgwick did not "hone himself" against the hard rocks of North Wales in order to store up ammunition for Darwin.

A second criticism leveled against the standard geologic column and scale of time is that the column simply does not exist except in the minds of geologists. In a trivial sense this is true. At no place on earth has an uninterrupted sequence of strata been found ranging in age from Cambrian to yesterday. Universal formations do not exist: the task of the historical geologist is to piece back together whatever parts of rock sequences have escaped destruction by earth movements and erosion. The game is that of a three-dimensional jigsaw puzzle involving also the fourth dimension of time. "Standard stratigraphic column" and "geological time scale" are conceptual terms, based on observations and inferences from observations. Geology, like other sciences, is full of theoretical terms. Stratum, incidentally, is one such. Though everybody agrees that strata exist, no one has seen one—only tops and bottoms and cross-sections, from which the mind's eye reconstructs the whole.

CHAPTER ELEVEN

A Question of Tempo

The outcome of any serious research can only be to make two questions grow where only one grew before.

THORSTEIN VEBLEN

In the early years of the nineteenth century, geology emerged as a science in its own right, a subject taught in leading universities, widely publicized in journals, and increasingly pursued by organized groups. The opposing views of the Huttonian plutonists and the Wernerian neptunists were championed by two members of the faculty at the University of Edinburgh. In 1802 John Playfair published his lucid *Illustrations of the Huttonian Theory of the Earth*, and seven years later his colleague, Robert Jameson, responded with his *Elements of Geognosy* (which has recently been reprinted under the more informative title, *The Wernerian Theory of the Neptunian Origin of Rocks*). Evidently some readers found Jameson's *Elements* a little over their heads. In any case Robert Bakewell seized the opportunity to write a textbook for readers unskilled in the classification of rocks and minerals. His *Introduction to Geology*, first published in 1813, explains that geognosy is no more than a fancy name for geology, and goes on to define a geognost as "a perfect disciple of Werner who has lost the use of his own eyes by constantly looking through the eyes of his master" (*1*).

Scientific societies specializing in geology arose in Britain alongside the Royal Societies of London and Edinburgh. The Geological Society of London was organized in 1807, and in the following year the Wernerian Natural History Society split from the Royal Society of Edinburgh under the leadership of Jameson. In 1810 geology was at last cited as a separate topic of reference in the fourth edition of the *Encyclopaedia Britannica*.

The two societies at Edinburgh and London, though both organized for the advancement of knowledge about the earth, followed different paths to that end. The Wernerians at Edinburgh were mainly bent on marshalling evidence only in support of the neptunian theory. The Londoners, by contrast, professed to be in search of hard facts. In the first volume of the Geological Society's *Transactions*, published in 1811, there was printed a motto taken from Bacon's *Novum Organum* which advised natural philosophers to cultivate "clear and demonstrable knowledge instead of attractive and probable theory" (*2*). The contents of this volume adhere to the spirit of the motto, consisting as they do of factual accounts of local structural geology, rocks, minerals, and a chalybeate spring on the Isle of Wight. Whewell aptly characterized these bare-fact communications as long and dreary. But, as a modern philosopher of science has observed, "facts" often turn out to be small theories. Inevitably the collection of hard facts by the rapidly expanding membership of the Geological Society of London led to a clash of minds concerning how these fit together in broader frameworks. The result: a debate between the uniformitarians and the catastrophists, started around 1830 and not yet entirely ended.

Members of both sides agreed that the surface of the earth has undergone repeated changes in configuration in the course of time. Among other matters, they disagreed with regard to the tempo of change. Catastrophists held that the prevailing tranquillity of nature has been interrupted from time to time by convulsive releases of energy which caused revolutionary changes both in geography and in the populations of organisms. The uniformitarians, on the other hand, maintained that past changes in configuration have occurred at about the same rate we see today.

Catastrophism was the prevailing doctrine during the early nineteenth century, but its origins are found in geological writings much older. Steno's explanation of the origin of lowlands in Tuscany by collapse of underground hollows eaten out "by the force of fire and water" was a catastrophist hypothesis. Hooke appealed to the violent release of energy during earthquakes and volcanic eruptions to turn plains into mountains and vice versa. Burnet attributed the irregularities of the crust to its collapse and flooding by the waters of the Deluge. The catastrophic effects of Noah's Flood also figured

Georges Cuvier (Portrait from the DeGolyer Western Collection, Southern Methodist University)

prominently in the theories of Woodward and Whiston. Buffon accounted for the diminution of the sea by collapse of crustal "blisters." Hutton thought that the continents were catastrophically uplifted from the sea floor, and he allowed for the possibility that they could as suddenly collapse and founder.

Georges Cuvier was, however, the person whose writings gave greatest credence to catastrophism. He first set forth his theory of the earth in 1811, and shortly afterward Jameson translated the text into English (*3*). Cuvier used paleontological evidence to show that the continents had been repeatedly flooded by the ocean, but he went further to insist that these advances and retreats were convulsive and lethal.

> These repeated irruptions and retreats of the sea have neither been slow nor gradual; most of the catastrophes which have occasioned them have been sudden; and this is easily proved, especially with regard to the

> last of them, the traces of which are most conspicuous. In the northern regions it has left the carcases (*sic*) of some large quadrupeds which the ice had arrested, and which are preserved even to the present day with their skin, their hair, and their flesh. If they had not been frozen as soon as killed they must quickly have been decomposed by putrefaction. But this eternal frost could not have taken possession of the regions which these animals inhabited except by the same cause which destroyed them; this cause, therefore, must have been as sudden as its effect. The breaking to pieces and overturnings of the strata, which happened in former catastrophes, shew plainly enough that they were sudden and violent like the last; and the heaps of *debris* and rounded pebbles which are found in various places among the solid strata, demonstrate the vast force of the motions excited in the mass of waters by these overturnings. Life, therefore, has been often disturbed on this earth by terrible events—calamities which, at their commencement, have perhaps moved and overturned to a great depth the entire outer crust of the globe, but which, since these first commotions, have uniformly acted at a less depth and less generally. Numberless living beings have been the victims of these catastrophes; some have been destroyed by sudden inundations, others have been laid dry in consequence of the bottom of the seas being instantaneously elevated. Their races even have become extinct, and have left no memorial of them except some small fragment which the naturalist can scarcely recognize. (*4*)

Cuvier's theory was based on three principal lines of evidence, the first of which related to fossils. His studies of Cretaceous, Tertiary, and Quaternary sequences around Paris showed that the bones of quadrupedal animals belonging to species now living are found only in the most recent of these deposits. At lower levels in the stratigraphic column are found remains of extinct species of elephant, rhinoceros, hippopotamus, and mastodon belonging to known genera or closely allied to them. At still lower horizons the bones belong to genera not extant, and around the base of the Tertiary bones of mammals are missing or very rare. This sequence clearly indicates that during the later epochs of geological history whole populations of quadrupedal animals have been exterminated. Since the strata in the Paris Basin are essentially flat-lying this extermination must have been due not to crustal dislocation but to recurrent flooding, as evidenced by strata containing shells of marine organisms separating layers containing bones of the terrestrial quadrupeds.

Cuvier's second line of evidence was based on structural considerations. The foothills of the great mountain chains are mostly formed of stratified rocks, but unlike the flat-lying beds that floor the lowest and most level parts of the earth, these strata are inclined and otherwise deformed. The deformed strata underlie the horizontal beds of the lowlands and therefore must be the older; moreover, the fossils they contain are unlike those found in the beds above, and in most instances belong to extinct species. Thus the rocks of the foothills bear witness to ancient revolutions accompanied by deformation of the crust and destruction of many different kinds of organisms. The unfossiliferous rocks forming the highest mountains show by their contorted and fractured character that revolutions occurred in the early stages of earth history before the existence of living creatures. (*5*).

Lastly, Cuvier noted as evidence for violent revolutions the occurrence of "numerous and prodigiously large blocks of primitive substances scattered over the surfaces of the secondary strata, and separated by deep vallies from the peaks or ridges whence these blocks must have been derived" (*6*). His reference here is to the exotic boulders of the kind Playfair had called "erratics"; for example, those scattered over the southern foothills of the Jura Mountains which by their lithology must have been moved far northward from sources in the Alps. Cuvier concluded that these masses must either have been tossed to their present positions by eruptions, or else somehow have been transported downhill from their roots in the southern mountains. In the latter instance the valleys intervening between source and present location of the erratics must have been incised since the event of overland transportation. Either way, something extraordinarily energetic, something revolutionary, is indicated.

Cuvier was well aware that the configuration of the earth's surface is constantly changing. He describes in considerable detail the causes of these changes, as related to mass-wasting manifested in landslides and avalanches, to the day-to-day work of wind and waves, and to volcanic activity. But he concluded that these agencies, taken singly or all together, are insufficient sources of energy for the elevation of great mountain ranges.

Cuvier believed that the last great revolution was witnessed by humans.

> . . . if there is any circumstance thoroughly established in geology, it is, that the crust of our globe has been subjected to a great and sudden revolution, the epoch of which cannot be dated much farther back than five or six thousand years ago; that this revolution had buried all the coun-

> tries which were before inhabited by men and by the other animals that are now best known; that the same revolution had laid dry the bed of the last ocean, which now forms all the countries at present inhabited; that the small number of individuals of men and other animals that escaped from the effects of that great revolution, have since propagated and spread over the lands then newly laid dry; and consequently, that the human race has only resumed a progressive state of improvement since that epoch, by forming established societies, raising monuments, collecting facts, and constructing systems of science and of learning. (*7*)

This strikingly original variation on a theme by Moses reflects Cuvier's belief that the Hebraic version of the Flood is only one of several traditions bearing witness to the fact that the crust of the earth was subjected to "a great and sudden revolution" some 6,000 years ago. He cites references from ancient Chaldean, Hindu, Persian, and Chinese records to prove his point, and mentions Humboldt's suggestion that references to a deluge may also be found in the "barbarous hieroglyphics" of the American Indians (*8*). Moreover, he insisted that Noah's Flood was not a unique event, but only the most recent of several great "irruptions of the sea." Records of earlier floods are recorded in discontinuities within the fossil record.

In closing his essay, Cuvier made a strong appeal for studies of the *sequences* in which different kinds of fossils naturally occur in the rocks. In his view, this would represent a great accomplishment, for "man to whom only a short space of time is allotted upon the earth, would have the glory of restoring the history of thousands of ages which preceded the existence of the race, and of thousands of animals that never were contemporaneous with his species" (*9*).

"Thousands of ages" is about as close as Cuvier came to estimating the duration of geologic time, but it was enough to draw the fire of his translator. In some 150 pages of notes which Jameson appended to the *Essay*, he set Cuvier straight on several points. The high inclinations commonly seen in strata of the great mountain ranges are original and not the result of crustal dislocations, and the erratic blocks found in the Juras and other parts of Europe were transported from their sources by the force of water (i.e. the Flood). Finally, and emphatically, "our continents are not of a more remote *antiquity* than has been assigned to them by the sacred historian in the book of Genesis, from the great era of the Deluge" (*10*).

In England the debate between the uniformitarians and cata-

Charles Lyell (Portrait courtesy of the Geological Society of London)

strophists was triggered by the publication in 1830 of Charles Lyell's *Principles of Geology*. Among the many remarkable features of this work is the fact that it got written in the first place, for had the author's eyesight been better it is not unlikely that he might have spent his life in the practice of law (*11*).

Lyell was born in 1797—the year of Hutton's death—at Kinnordy, the family home at the foot of the Grampian Mountains in eastern Scotland. Soon after his birth the family relocated in southern England near the fashionable town of Southampton. Rents from the estate in Scotland provided a comfortable living, and the leisure for the senior Lyell to pursue his intellectual interests in botany and literature. The father's concern for natural history was soon reflected in the son's hobby of collecting butterflies and aquatic insects. However, the senior Lyell intended that his namesake and eldest child should attain financial independence practicing law. As a first step toward this goal, Charles was enrolled at Exeter College, Oxford, in 1816. Prior to his graduation in 1819 his interest in geology was first aroused by a reading of Bakewell's popular text, and then heightened by the Reverend William Buckland's geological lectures and occasional field excursions. Also, in the summer of 1818 he had taken a lengthy trip with his family through France, around Switzerland, and across the Alps into Italy, carefully observing the varied geological features all along the way.

Protracted sessions of reading for his examinations at Oxford had produced a painful inflammation of the eyes. Though this was an ill omen for a young man preparing for a profession which by its nature requires much reading, Mr. Lyell nevertheless enrolled his son at Lincoln's Inn in London for training in the technicalities of English law. Despite recurring problems with his vision, Lyell applied himself with a dogged diligence to his studies, and in 1822 was called to the bar.

Meanwhile Lyell continued to pursue geology as an avocation. In 1819, the year of his graduation from Oxford, he had been elected to fellowship in the Geological Society of London. At every possible opportunity he took excursions into the field, especially into southeastern England and the Isle of Wight. In 1823 he was elected one of the two secretaries of the Geological Society, and thus assumed a responsibility for reading and editing papers to be published in the *Transactions.* That same year he visited Paris, where he was able to establish acquaintance with Cuvier, Humboldt, and Brongniart. There Louis Constant Prevost, student of Brongniart and distinguished geologist in his own right, escorted Lyell on field trips in the Paris Basin.

In 1825 Lyell was still being urged by his father to practice law, evidently as a means of providing financial security for the young man's unmarried sisters. By this time he would have much preferred a career in geology, but hesitated to make a final decision because he could see no prospects of making a living, much less a fortune, at this pursuit.

This dilemma was resolved when Lyell discovered that he could write for money. Late in 1824 John Coleridge became editor of the *Quarterly Review.* Coleridge, who was also a fellow of Exeter College, and who wanted to see more writing on science and medicine in the *Review,* invited Lyell in 1825 to prepare articles for his journal—for pay. Within two years Lyell produced five articles. One of these, published in 1826, was a lengthy review of the Geological Society's *Transactions;* but in this essay he went far beyond the specific topic at hand to summarize the state of geology generally. If Lyell could not make a living at the practice of geology, it now seemed possible to do so through sale of his writings on the subject.

Lyell probably began writing his *Principles* late in 1827. During the next two years he gathered additional information in the course of field studies in France and Italy. In the company of Murchison he made a traverse from Calais to Marseilles, stopping in the Auvergne District for detailed studies of volcanic deposits and freshwater limestone. Then around the coast and into Italy and along the northern side of the Po Valley to Venice, at which point Murchison left the party. Lyell continued his journey southward to Sicily, spending considerable time in studying the effects of volcanic activity around the Bay of Naples and at Mount Etna, and gathering new information on the sequence of Tertiary strata.

The *Principles* proved to be an eminently successful and influential work. After the first volume was issued in 1830, second and third volumes appeared respectively in 1831 and 1833. A part of this success lay in the fact that Lyell had not only synthesized the geological knowledge of his day around a coherent theory, but also offered in support of his views many fresh observations based on his own field studies. His literary style combined grace with clarity, so that readers without technical training could understand his message. The fact that the book excited controversy only added to its popularity. Moreover, Lyell did not let his work get out of date; he constantly made changes in the text to keep pace with the increase of knowledge about the earth, and when he died in 1875 he was busy revising the twelfth edition.

Lyell's uniformitarian system was based upon Hutton's theory, but with several amendments. Like Hutton, he excluded from geology all questions as to the origin of things, and sought to explain the former changes of the earth's crust as the work of natural agents. This position he highlighted in the subtitle of his book, which states that the aim here is "to explain the former changes of the earth's surface by reference to causes now in operation." He accepted Hutton's conclusion that granite is an intrusive igneous rock, hence younger than the rocks it has invaded and not a sediment forming a primitive and universal substratum. And, most important of all, he concurred with Hutton's reasoning that if all past changes on the globe have been brought about by the slow agency of existing causes, then the earth must be ancient beyond human powers of comprehension.

On the other hand, Lyell found Hutton's theory deficient in underestimating the significance of fossils in the ordering of past events. Also he rejected the propositions that subterranean heat is necessary for the consolidation of marine sediments, and that new continents may have been elevated from the floor of the sea by violent and paroxysmal convulsions (*12*).

Looking back on the history of geology, Lyell identified the principal sources of prejudice which have retarded the progress of the science. These relate to preconceptions regarding the duration of past time, and general ignorance of what goes on in the depths of the earth and beneath the waters of the world ocean.

Little wonder, Lyell explains, that many of the early philosophers conceived that the lands are like ruins. "As dwellers on the land, we inhabit only about a fourth part of the surface; and that portion is almost exclusively a theatre of decay, and not of reproduction." (*13*) We know that the wearing down of the continents by erosion is necessarily accompanied by the building of new strata on the floors of lakes and seas. But we are not witnesses to this constructive phase of natural operations.

> He who has observed the quarrying of stone . . . and has seen it shipped for some distant port, and then endeavours to conceive what kind of edifice will be raised by the materials is in the same predicament as a geologist, who, while he is confined to the land, sees the decomposition of rocks, and the transportation of matter by rivers to the seas, and then endeavours to picture to himself the new strata which Nature is building beneath the waters. (*14*)

Geology might have progressed more rapidly, Lyell submits, if mankind were amphibious, and could observe directly the chain of events leading from the weathering of rocks in the atmosphere to deposition on the bottom of the sea of the sediment so produced. Amphibious geologists could compare directly the products of volcanic action poured out upon the land with those erupted over the sea floor. Even if men had amphibious capabilities, Lyell reasons, their ignorance of changes taking place beneath the rocky surface of the solid earth would remain abysmal. If some sooty sprite be imagined to have the capabilities of moving about in the confines of the solid interior, he might observe, as we cannot, the upward movement of molten rock toward the surface, might see how the "roof" of fossiliferous strata is ingested or fused and recrystallized—and then might get the whole story wrong. The fossiliferous strata, for which no analogues could be found in the subterranean world, might reasonably be interpreted as remnants of a substance that once formed the entire globe, and since has been reduced to a discontinuous rind as the earth progresses toward a molten state. The serious point that Lyell so playfully makes is of course that prejudices naturally arise from limitations in experience, "resulting from the continual contemplation of one class of phenomena to the exclusion of another" (*15*).

Of the several impediments to the advance of geologic knowledge, Lyell gave most attention to preconceptions with regard to the duration of past time. The gist of his argument here is that given a series of connected events arranged in chronologic order, the relating of these in an historical format will make sense or not depending on the time allowed to have lapsed between the first and last in the series. Duration sets the tempo.

> How fatal every error as to the quantity of time must prove to the introduction of rational views concerning the state of things in former ages, may be conceived by supposing the annals of the civil and military transactions of a great nation to be perused under the impression that they occurred in a period of one hundred instead of two thousand years. Such a portion of history would immediately assume the air of a romance; the events would seem devoid of credibility, and inconsistent with the present course of human affairs. A crowd of incidents would follow each other in quick succession. Armies and fleets would appear to be assembled only to be destroyed, and cities built merely to fall in ruins. There would be the most violent transitions from foreign or intestine war to periods of profound peace, and the works effected during the years of disorder or tranquillity would appear alike superhuman in magnitude. (*16*)

In explaining why wrong ideas on the duration of past time have obstructed the development of geologic history, Lyell avoided a head-on attack against adherents to the Biblical chronology. Rather, he professed sympathy toward "men of great talent and sound judgment" who in earlier centuries had produced fantastical theories of the earth. Considering their unshakable faith in the proposition "that the earth was never the abode of living beings until the creation of the present continents, and of the species now existing," their theories were bound to be extravagant. Here again Lyell uses historic analogy to make his point. Suppose, he suggests, that Champollion and other pioneer Egyptologists had arrived in Egypt under the delusion that the banks of the Nile had never been populated by the human race before the beginning of the nineteenth century.

> The sight of the pyramids, obelisks, colossal statues, and ruined temples, would fill them with such astonishment, that for a time they would be as men spell-bound—wholly incapable of reasoning with sobriety. They might incline at first to refer the construction of such stupendous works to some superhuman powers of a primeval world. (*17*)

Discovery of the mummies would set off new trains of speculation. Perhaps these objects did not belong to humans, but are sports of nature, generated in the ground by some plastic virtue. Or possibly these are seeds of individuals yet to be born. "These speculations, if advocated by eloquent writers, would not fail to attract many zealous votaries, for they would relieve men from the painful necessity of renouncing preconceived opinions" (*18*).

If we were convinced that the Great Pyramid was raised in one day, we would be justified in ascribing its erection to some superhuman power. By the same token, Lyell insists, we must appeal to some extraordinary and revolutionary cause if we conceive that a great mountain range or a whole continent has been elevated in a short period of time. Take the Andes, for example.

> We know that during one earthquake the coast of Chile may be raised for a hundred miles to the average height of about three feet. A repetition of two thousand shocks, of equal violence, might produce a mountain chain one hundred miles long, and six thousand feet high. Now, should one or two only of these convulsions happen in a century, it would be consistent with the order of events experienced by the Chilians from the earliest times: but if the whole of them were to occur in the next hundred years, the entire district must be depopulated, scarcely any animals or plants could survive, and the surface would be one confused heap of ruin and desolation. (*19*)

Most of the first volume of the *Principles* and all of the second describe geological processes now in action, to the end of demonstrating that operating at present intensity these are adequate to account for all the geological phenomena of the past, given eons of time. Writing from his own experience in Sicily, he recites evidence to show that Mount Etna is made of innumerable lava flows, no one more voluminous than those witnessed during the intermittent outpourings of historic times. Thus the mountain must have been built up gradually over a span of time immense by human standards. Lyell took special pains in offering gradualistic alternatives to phenomena cited by the catastrophists as evidence for violence. The erratic blocks that cover much of northern Europe are not relics of frightful explosions, nor the sediment of powerful torrents unlike any known from experience, but were dropped from melting icebergs floating over the continents during the most recent of many encroachments by the sea. Catastrophists had pointed to mixtures of terrestrial and marine fossils in certain strata as proof of devastating revolutions of the kind envisioned by Cuvier. Lyell insisted that strata containing such mixtures are forming today in the delta of the Mississippi. Here the waters of annual floods regularly carry parts of terrestrial plants and bones of terrestrial animals into lagoons or shallow waters offshore, where they are buried on the bottoms along with the skeletons of marine fish and shellfish.

> Yet many geologists, when they behold the spoils of the land heaped in successive strata, and blended confusedly with the remains of fishes, or interspersed with broken shells and corals, imagine that they are viewing the signs of a turbulent instead of a tranquil state of the planet. They read in such phenomena the proof of chaotic disorder, and reiterated catastrophes, instead of indications of a surface as habitable as the most delicious and fertile districts now tenanted by man. They are not content with disregarding the present course of nature . . . but they often draw conclusions concerning the former state of things directly the reverse of those to which a fair induction from facts would infallibly lead them. (*20*)

Lyell's basic principles were twofold, though sometimes he lumped them together as a single proposition. For example, in an address before the Geological Society of London, delivered in 1851, he asserted that "the ancient changes of the animate and inanimate world, of which we find memorials in the earth's crust, may be similar both in kind and degree to those which are now in progress" (*21*). In private correspondence Lyell could better reveal the depth of his conviction. A letter written to Murchison shortly before the first volume of the *Principles* was published contains these statements. "No causes whatever have from the earliest time to which we can look back, to the present, ever acted, but those now acting," and, "they never acted with different degrees of energy from that which they now exert" (*22*). The first of these propositions ruled out supernatural and fanciful causes. The second, which obviously doesn't necessarily follow from the first, was aimed against those catastrophists who argued that the earth has constantly been losing energy since its (probably fiery) beginning, so that the forces of change must have operated more vigorously in the past than now.

Less than a year after the first volume of the *Principles* was published, the Reverend Adam Sedgwick devoted most of his attention to the book in his address as retiring president of the Geological Society. He praised Lyell for having taught him so much about "geological dynamics," and professed to agree with "nineteen-twentieths of his work." As for the remaining twentieth part to which he took exception, that included both of Lyell's basic principles.

Sedgwick regarded as gratuitous any assumption that the only causes of change that have operated during the past are those now in operation. "What is this," he asks, "but to limit the riches of the kingdom of nature by the poverty of our own knowledge?" Granted that the "primary laws of matter" are immutable, it does not follow that the operations of these laws will always produce the same end results.

> I believe that the law of gravitation, the laws of atomic affinity, and, in a word, all the primary modes of material action, are as immutable as the attributes of that Being from whose will they derive their only energy . . . but (these) very powers themselves act under such endless modification, sometimes combined together, and sometimes in conflict, that there follow from them results of indefinite complexity. . . . (*23*)

Compelling evidence for *change* in the course of time is to be found in the fossil record, Sedgwick emphasized. Though he regarded the idea that one species could change into another as no less "monstrous" than the doctrine that organisms spring spontaneously from mud, he perceived a definite *progression* from lower to higher forms of life in the succession of geological systems. Remains of mammals, he noted, are missing from the Paleozoic and very rare in the Mesozoic strata. The warm-blooded quadrupeds became increasingly abundant during the Tertiary Period, and man appeared late in the Quaternary. By what operation of natural laws, manifest in secondary causes at present in action, can we explain these changes in the organic world? The recent appearance of man is in itself "absolutely subversive of the first principles of the Huttonian hypothesis" (*24*).

Sedgwick found Lyell's second principle, relating to approximate constancy in energy of existing causes, no less gratuitous than the first. On theoretical grounds, the globular shape of the earth and the crystalline state of the oldest rocks in the crust suggest that the planet was originally molten. In which case the earth's surface must have undergone a gradual refrigeration with time, and at least in the theater of life the thermal forces that power Hutton's heat machine must have become less energetic. In this context it seems strange to assume that volcanic forces have always acted at about the same intensity as at present.

On empirical grounds, Lyell's second principle seemed to Sedgwick even less tenable. The broken and folded strata in the great mountain ranges, the discontinuities in the fossil record, and the erratic blocks scattered over northern Europe all argue that "paroxysms of internal energy, accompanied by the elevation of mountain chains, and followed by mighty waves desolating whole regions of the earth, were a part of the mechanism of nature" (*25*).

Sedgwick offers a catastrophist alternative to Lyell's hypothesis that erratics have been rafted from their sources by drifting masses of ice. Suppose that during a period of intense volcanic violence, Scandinavia were suddenly elevated. The resulting outrush of water would then provide a force sufficient to transport these great boulders southward over the plains of northern Europe. And yet, he complains,

> Mr. Lyell will admit no greater paroxysms than we ourselves have witnessed—no periods of feverish spasmodic energy, during which the very framework of nature has been torn asunder. The utmost movements that he allows are a slight quivering of her muscular integuments. (*26*)

Sedgwick ends his address with an apology and a profession of faith. He confesses his error in assuming that all the superficial gravel of the earth is the product of one violent and transitory event. "We ought, indeed, to have paused before we first adopted the diluvial theory, and referred all our old superficial gravel to the action of the Mosaic flood" (*27*). Perhaps future discoveries of the diagnostic human remains and artifacts will tell which parts of these deposits are contemporary with Noah. Or, possibly, it was not intended that these remains should be preserved, or found. In any case the evidence for prehistoric catastrophes of the kind recorded by Moses is compelling; therefore we cannot deny the possibility that what has happened in the distant past happened once during the few thousand years man has lived on earth.

Granted that geologists have their differences, Sedgwick concludes that "geology lends a great and unexpected aid to the doctrine of final causes. . . . I believe too firmly in the immutable attributes of that Being, in whom all truth, of whatever kind, finds its proper resting place, to think that the principles of physical and moral truth can ever be in lasting collision" (*28*).

Sedgwick's critique was only one of the earlier among many evaluations of the pros and cons of Lyell's system. In the years that followed first publication of the *Principles,* virtually every geologist of note expressed his opinion of this work. Lyell's accomplishment in synthesizing geological information from an amazingly broad variety of sources and geographical situations was acknowledged. Few denied that he had sorted out, as well as if not better than anyone before him, the kinds of changes taking place in the physical world at present, and had in most instances correctly related these changes to the material agencies responsible for them. But there were serious matters of disagreement, both formal and substantive, touched upon by Sedgwick and elaborated by others.

On logical grounds, dissenters objected to the *a priori* reasoning underlying Lyell's two guiding principles. If the earth is so old as Hutton and Lyell claimed, how can one assume that the agencies of change not only have been forever the same in kind but also have maintained the same average levels of energy that we witness today? Substantively, the principal issue lay in paleontologic evidence affirming that different kinds of organisms had inhabited the earth at different times, a phenomenon that Lyell couldn't account for by "causes presently in operation." Contorted and broken strata of the high mountain ranges, no less than the widely dispersed erratic blocks and continental gravels, seemed to call for forces of magnitude far greater than those released during recorded earthquakes and floods.

As the debate progressed, the issue of the earth's antiquity tended to fade. Sedgwick was willing to concede that geologists were entitled to equal time with the astronomers.

> . . . in the phaenomena of geology we are carried back . . . into times unlimited by narrow measures of our own, and we exhibit and arrange the monuments of former revolutions, requiring for their accomplishment perhaps all the secular periods of astronomy. (*29*)

Antiquity granted, William Whewell, in his review of the *Principles*, nevertheless twitted Lyell for taking "eternity for his working time."

> . . . Mr. Lyell seems to thirst for an antiquity of the earth even greater than that which is indicated by geological phenomena themselves. When he maintains, after Hutton, that we see in geology, as in astronomy, "no mark, either of the commencement or of the termination of the present order;" when he implies, that the strata seem to tell us the story of a perpetual recurrence of cycles of change of the same kind; he appears to forget that the geological series, long and mysterious as it is, has still a beginning. (*30*)

Causes of change aside, the uniformitarians and catastrophists were in general agreement as to the temporal sequence of major events in earth history, as the record was then known. Both sides accepted superposition and the other ordering principles that provide "before and after" relationships in historical geology. But temporal successions of events, no more than mere succession of notes in a musical composition, won't tell the rhythm of the piece. Given the then prevailing latitude in marking the score, the beat could be metronomic and the themes periodically recurring, as Hutton and Lyell wanted; or the beat could be variable and the configurations of movement nonrecurring, as nearly everybody else preferred. *Tempo giusto*, or *tempo rubato?*

CHAPTER TWELVE

A Plenitude of Events

In science the credit goes to the man who convinces the world, not to the man to whom the idea first occurred.

SIR WILLIAM OSLER

With the progress of geological investigations in western Europe during the first half of the nineteenth century, the number of events that had to be fitted into the chronicle of the earth's history increased enormously. Studies of the rocks and landforms produced by volcanic activity disclosed that volcanism is no recent and insignificant phenomenon, as the Wernerians claimed, but rather a persistent and potent agency of change intermittently active throughout geologic time. Studies of glaciers and of the debris transported by them led to the conclusion that much of the northern hemisphere was formerly covered with ice. The new sciences of volcanology and glaciology not only swelled the record of past events but also interpolated events of a spectacular kind that increased popular interest in geology. Neptunism faded as insurmountable evidence accumulated to prove that layers of basalt deep down in the geological column were ancient lavas poured out upon the surface of the sold earth, and not precipitates laid down on the bottom of a once universal ocean. Diluvialism lost all scientific credence as the erratic boulders and other surficial debris ascribed to Noah's Flood were recognized as the work of glaciers. One might say that the first ism got burned up and the second frozen out.

Curiously enough, the area that provided the critical evidence which led to the demise of neptunism is a broad plateau in south-central France where there has not been a volcanic eruption in historic times. This Central Plateau (Massif Central) borders the western side of the Rhone Valley, but drains mostly northward into the Loire. Granite and other crystalline rock (the Primitive Rocks of the neptunists) form the foundation over most of the area. On the average, the plateau stands about 3,000 feet above sea level, but in the Auvergne highlands around the city of Clermont Ferrand mountains rise to twice this elevation. Here hills and mountains formed of lava and cinders appear in all stages of dissection by streams. The Chaîne des Puys consists of almost a hundred small cones (puys) and associated lava flows, many as perfectly preserved as any found in areas of active volcanism. Larger accumulations of volcanic materials, as found in the Cantal and Mount Dore, are so extensively dissected that the original landforms must be inferred from their ruins (*1*).

The volcanic origin of the puys was recognized in the mid-eighteenth century by Jean-Étienne Guettard (1715–1786), keeper of the duc d'Orléans' natural history collection. Returning from a visit to Vesuvius and nearby volcanic fields around the Bay of Naples, Guettard visited the Auvergne in 1751 and instantly recognized the cinder cones for what they are. His memoir announcing "the volcanic origin of certain mountains in France," published the following year, found little favor among local savants who preferred to believe that the cinder cones were slag heaps from iron furnaces worked by the Romans.

Nicolas Desmarest (1725–1815) was no more successful than Guettard in overturning neptunian notions concerning the origin of basalt. A French civil servant whose principal duty was to improve manufacturing techniques, Desmarest entered the Auvergne first in 1763. At intervals over the next eleven years he prepared maps showing the areal relationships of the basalt bodies to each other and to the basement of crystalline rocks. Some of the younger flows he could trace to their points of issuance in the craters or at the bases of well preserved cones. Where the sources of the basalt could not be determined, he found that the ground beneath had been hardened as though baked. His memoirs on the volcanic origin of the basalt in the Auvergne, published in 1771, were ignored in most quarters. To those who contested his views his reply was, "Go to the Auvergne and see." Several of Werner's students did indeed follow this injunction, and their faith in the Wernerian system was shaken in varying degrees proportional to the time they spent on the Massif Central. Their reports of misgivings did not, however, move Werner to change his views. If volcanos were allowed to have erupted through *Primitive* crystalline rocks, then the whole neptunist scheme of orderly and universal formations would topple.

One of the half-dozen or so persons whose collective findings at last put an end to the neptunist doctrine was George Julius Poulett Thomson (1797–1876). Volcanos fascinated Thomson: when he was only twenty he embarked on the first of three consecutive winters of studying the volcanos of Italy, Sicily, and the Lipari Islands. He returned then to Cambridge, where Sedgwick was beginning his career as professor of geology. At that time Sedgwick was not the best advisor for a young enthusiast of volcanism, for, as he recalled in a letter to Lyell, he was "eaten up with Wernerian notions, ready to sacrifice myself to that creed—a Wernerian slave" (*2*). But Thomson had an independent mind, and resolved to give volcanism its due place in geology. After his marriage in 1821 to the daughter of William Scrope, earl of Wiltshire, he also had the independent means to pursue that goal. Following the marriage he changed his surname to Scrope, and it is as Scrope or sometimes Poulett-Scrope that he is cited in the histories and encyclopedias.

Soon after his wedding Scrope took up headquarters at Clermont and spent several months studying the extinct volcanos of the Auvergne. He then went on to Italy, where he had the good fortune to witness the Vesuvian eruption of 1822. The results of his studies appeared in 1825, in a book entitled *Considerations on Volcanos,* which has been called the first systematic work on volcanology. Further studies in the Auvergne were followed by the publication in 1827 of his *Geology and Extinct Volcanos of Central France*, a second and enlarged edition of which appeared in 1858.

Scrope's publications on the Auvergne reconfirmed and amplified the earlier findings of Guettard and Desmarest regarding the volcanic origin of the cones and basalt flows, provided conclusive evidence that the volcanism was an intermittent process extending over a long period of time, and proved the efficacy of stream erosion in modifying or destroying volcanic landforms. Over and above his contributions to volcanology, his work constituted a three-pronged attack against prevailing doctrines of catastrophism, neptunism, and diluvialism.

The success of Scrope's books on the extinct volcanos of the Auvergne derived not only from the clarity and simplicity of the text and the meticulous detail in which the volcanic rocks and landforms

George Poulett Scrope (Portrait courtesy of the Geological Society of London)

are described, but also from well-executed panoramic views of the landscape. Any reader who had seen a cinder cone in a field where volcanos are now active would find familiar shapes in Scrope's fine illustrations of the cratered hills along the Chaîne des Puys.

Moreover, Scrope could demonstrate that the flows radiating from the cinder cones were identical in composition to layers in more dissected areas where basalt forms the caps of mesas and in some places is interbedded with deposits laid down in ancient lakes. Where some of the flows extend along present stream valleys, the stream has incised a new valley along the side of the flow, running along a bed which now lies several hundred feet below the original one at the base of the lava. This latter circumstance prompted Scrope to ask a question of those diluvialists who would ascribe the contours of the present landscape to the work of Noah's Flood.

> Now, if the first excavation of these valleys is to be accounted for by . . . a deluge, to what are we to attribute the second . . . ? Not, most certainly, to a *second deluge;* for the undisturbed condition of the volcanic cones, consisting of loose scoriae and ashes, which actually let the foot sink ankle-deep in them, forbids the possibility of supposing any great wave or debacle to have swept over the country since the production of these cones. (*3*)

On purely philosophic grounds Scrope was opposed to invoking catastrophes of any kind, watery or otherwise, to account for geological phenomena.

> As the idea imparted by the terms cataclysm, catastrophe, or revolution, is extremely vague, and may comprehend anything you choose to imagine, it answers for the time as an explanation; that is, it stops further inquiry. But it has the disadvantage of stopping also the advance of the science by involving it in obscurity and confusion. (*4*)

Volcanic cone and crater near the village of Jaujac in the Auvergne. The village is built on a lava flow that issued from a breach in the crater, and has since been eroded to form cliffs along the Alignon River. (From the frontispiece of Scrope's The geology and extinct volcanos of central France)

No need to invoke unusual causes of change, Scrope insisted, if those causes currently in action are sufficient to account for the effects under investigation—given sufficient time. Referring to the effects of erosion in modifying the volcanic landforms, Scrope reveals himself as a gradualist and uniformitarian.

> Even if it were allowable to have recourse to vague and hypothetical conjectures, we can conceive no *gradual and progressive excavating forces*, other than those which are still in operation wherever rains, frosts, floods, and atmospheric decomposition act upon the surface of the earth. To these agents then we must refer the effects in question, of which, with an unlimited allowance of *time*, no one will pronounce them to be incapable. (5)

In both of his earlier books Scrope emphasized that the careful study of geologic processes now in operation affords the only rational method of inquiry into the history of the earth's surface. He surmised that Lyell had "imbibed that philosophical conviction" before writing his *Principles of Geology*. In any case, both men recognized that if present episodic processes such as volcanism and generally slow-working processes such as stream erosion are to be invoked as agencies to raise and level mountains, vast periods of time must be assumed.

> The time that must be allowed for the production of effects of this magnitude by causes evidently so slow in their operation is indeed immense; but surely it would be absurd to urge this as an argument against the adoption of an explanation so unavoidably forced upon us. The periods which to our narrow apprehension . . . appear of incalculable duration, are in all probability but trifles in the calendar of Nature. It is Geology that, above all other sciences, makes us acquainted with this important though humiliating fact. Every step that we take in its pursuit forces us to make almost unlimited drafts upon antiquity. The leading idea which is present in all our researches, and which accompanies every fresh observation, the sound which to the student of Nature seems continually echoed from every part of her works, is—
>
> Time!—Time!—Time! (*6*)

More recent geological studies in the Auvergne have shown that the higher volcanic mountains there have not only been dissected by streams but also eroded by glacial ice. When the first edition of Scrope's work on the extinct volcanos appeared in 1827, however, the importance of glaciers in remodeling the landscape was not generally understood. The concept of an "Ice Age" did not in fact gain general acceptance until after 1840, the year when Louis Agassiz's classic *Études sur les Glaciers* was published. Like some other landmark works in geology, the *Études* were more the outcome of fortuitous circumstances than the product of design.

John Louis Rodolphe Agassiz was born in 1807 at the parsonage of Motier-en-Vuly, a Swiss village on the Lake of Morat (Murtensee). The Vuly is an agricultural area in the lake country fronting upon the Jura Mountains, along a stretch of some thirty miles between Biel and Yverdon. One of Agassiz's biographers has suggested that this setting played a role in shaping his scientific career.

> Born and educated in such a place as Motier, surrounded by water and marshes, with the Oberland always in full view in front, and the summit of the Jura in the rear, it is no wonder that Agassiz became an ichthyologist and a glacialist. Everything which met his eye, from infancy until manhood, seems to have awakened in him a curiosity to know his surroundings. It was as natural for him to take to the study of fishes and of glaciers as it is for sons of seamen to go to sea, or for "vignerons" (vine dressers) to go to the vineyard, or for "gauchos" to ride on the prairies of South America, or for the Arabs to cross the desert on camels. (7)

Obviously this is making too much of geography, and in Agassiz's case too little of native intelligence, parental guidance, and academic training. The record shows that Agassiz had indeed acquired during his childhood some habits of the naturalist. For example, he established an aquarium on the grounds of the parsonage, using as a basin the recess in an erratic boulder of granite. This early interest in natural history may not have been entirely spontaneous, for, like all other children of the parish, Louis was under the tutelage of his father, whose pastoral duties included instruction at the elementary level.

At age ten Agassiz embarked on an academic odyssey, first to academies at Biel and Lausanne, where he received training in the liberal arts and sciences, then on in succession to Zurich, Heidelberg, and Munich for studies of medicine. While he was at Munich his interests in natural history were stimulated by an unexpected turn of events. One of his professors, working on reports of explorations in the Amazon Basin, suffered an interruption of his project by the death of a colleague who was to describe a large collection of Brazilian fishes. Agassiz was offered and eagerly accepted the responsibility of classifying the specimens. Acting on the advice that any published results might be discounted in academic circles unless the

Louis Agassiz (Portrait courtesy of the DeGolyer Geological Library, Southern Methodist University)

author were cited as a Doctor of Philosophy, he promptly went to the University of Erlangen to remove this deficiency. A year later, when the monograph on Brazilian fishes was issued in 1829, the author was Dr. L. Agassiz. For good measure Agassiz stayed at Munich for nine additional months to complete work for the M.D.

The monograph on fishes, with its Latin text and ninety well-executed plates, is now a collector's item. Appropriately dedicated to Cuvier, leading ichthyologist of the day, this work established Agassiz as a competent naturalist when he was only twenty-two. Encouraged by this first accomplishment Agassiz lost no time in announcing plans for a second major project. In 1830 he published a prospectus for a comprehensive classification of the fresh-water fishes of central Europe. Before this flier was off the press, however, he had conceived yet another and grander program of study—this one to deal with the paleontology of fishes.

Agassiz had begun to collect materials on fossil fish before he had completed medical studies. Late in 1831 he travelled to Paris, hoping for access to paleontological collections there through acquaintance with Cuvier and others, and wishing also to secure some kind of appointment that would sustain him and his research. These hopes were rewarded beyond all expectations when, only a few days after his arrival, Cuvier invited Agassiz to take up quarters in his laboratory.

For some fifteen years prior to Agassiz's arrival in Paris, Cuvier had been gathering information for a major work on fossil fishes. After observing Agassiz at work, however, he was so impressed with the young man's enthusiasm for that branch of research, no less than with his capabilities as anatomist and taxonomist, that he turned over to him all his drawings and notes on the paleontology of fishes, without stipulating joint authorship of any publications that might result. This act of generosity was to be among the last of many credited to the great French scientist: Cuvier died of cholera—only five months after his first meeting with Agassiz—during the epidemic that devastated Paris in the spring of 1832. Suddenly set adrift without a patron, Agassiz left Paris the following year and settled in Neuchâtel as professor of natural history in the small college there.

At Neuchâtel Agassiz made the best of a bad situation. Fewer than a hundred students were enrolled at the college, and the physical facilities were minimal. He lectured in the courtroom at the City Hall, established a museum at the Orphans' Home, and in company with five other citizens founded the Neuchâtel Society of Natural Sciences. In 1833 he married the talented Cecile Braun of Carlsruhe, who despite her dislike for Switzerland in general and for Neuchâtel and the Neuchâtelois in particular also made the best of a bad situation by drawing pictures of fossil fishes for her husband.

In 1834 the first number of *Poissons Fossiles* appeared, and was instantly recognized by geologists and zoologists alike as a significant work. Accordingly, when Agassiz visited England later in the same year to collect more specimens, he was warmly received by Lyell, Buckland, and other notables. His field work yielded some two thousand specimens, and the Geological Society of London provided space for their preparation and examination.

Agassiz's monumental work on fossil fishes was published over a period of eleven years in five volumes illustrated by 400 colored folio plates. More than 1,700 species are described. In 1836, when the project was not yet in mid-course, the Geological Society of London awarded Agassiz the Wollaston Medal. In making the presentation,

Lyell took note of the "small pecuniary aid" that the Society had provided to accelerate publication. Agassiz was in his thirtieth year, the youngest to receive this honor.

Meanwhile a train of events was developing that would divert Agassiz's attention from fossils to glaciers (*8*). Some time prior to 1816 the Swiss mountaineer Jean-Pierre Perraudin made known to his friends his conviction that the large boulders scattered along the sides of the alpine valleys, high above the beds of present streams, had been transported and left there by glaciers (*9*). Evidently the only other person who was willing to entertain the idea at the time was the engineer Ignace Venetz. In 1821 Venetz read a paper before the Swiss naturalists in which he supported Perraudin, and suggested that the alpine glaciers are the remnants of former much more extensive masses of ice which formed the ridges of boulders (moraines) now far removed from the ends of present glaciers. This event of massive glaciation, he concluded, must be assigned to some epoch "which is lost in the darkness of time" (*10*). The paper went unnoticed, and the manuscript lay in the files of the Helvetic Society of Naturalists until Jean de Charpentier, director of the salt works at Bex, dug it out and had it published in 1833. In the interim Venetz had expanded his concept of the extent of former glaciation to envision a great sheet of ice extending from the Alps as far north as the Jura. His friend de Charpentier, who had at first been skeptical toward this idea, on examination of the evidence came to accept it. After the publication of Venetz's paper he felt free to speak to the subject himself, and so in 1834 addressed the Helvetic Society on glaciers as the most likely agents for the distribution of erratic blocks throughout Switzerland. Here de Charpentier carefully marshalled the evidence for past glaciation on the grand scale, as related to erratics, to moraines, and to polished and striated pavements. This report, widely disseminated through publication in several different journals, stirred up much discussion; but the general reaction among naturalists was one of disbelief, combined in some instances with scorn and mockery (*11*). Agassiz was an outspoken opponent.

De Charpentier was a mining engineer by profession and a naturalist by avocation. At the Freiberg School of Mines he was a classmate of Alexander von Humboldt and others who had since become reputable men of science. His home near Bex was a resort for intellectuals and a repository for collections in natural history. When Agassiz was still a student at Lausanne he met de Charpentier and developed a respect and fondness for the man. This friendship was long sustained, and in 1833 de Charpentier invited his young friend to visit him at Bex. But work, and the business of settling down to married life, delayed an acceptance until the summer of 1836.

What was intended as a summer vacation turned out to be a summer school in glacial geology for Agassiz. De Charpentier and Venetz showed Agassiz the field evidence for former glaciation in the Rhone Valley, the Valley of Chamonix, and other localities. In the course of a few weeks Agassiz's attitude toward the glacial hypothesis changed from disbelief to whole-hearted endorsement; and during the following fall he found additional evidence for glaciation in the Jura.

> Agassiz, with his extraordinary imagination, saw that the phenomenon of the extension of old glaciers had not been confined to the Rhone Valley, but must have been general, and formed a special period in the history of the earth. . . . In a word, Agassiz's sojourn at Bex, under the teaching of de Charpentier, had taught him, with his far-reaching thoughts, to add an entirely unexpected, and, at that time generally unacceptable, stage to the various periods which the earth had passed through; namely, the Ice-age. (*12*)

In July of 1837 the Helvetic Society met in Neuchâtel. Agassiz had been elected president, and thus was scheduled to deliver the opening address. Doubtless the members expected to hear something about fossil fishes. Instead Agassiz lectured on the glacial hypothesis, reading from a manuscript he had struck off the night before.

In his famous "Discours de Neuchâtel" Agassiz gave full credit to Venetz and de Charpentier for founding the glacial doctrine (*13*). He recited the evidence provided by moraines, polished and striated pavements, and erratics in support of their theses that glaciers have formerly been more extensive. That done, he turned to the underlying cause of continental glaciation.

Agassiz's theory of glaciation was based on two conceptions which were commonly held during the early nineteenth century. He subscribed to the belief that the temperature of the earth's surface has diminished in the course of time. Also he held that the history of life has been a record of alternate annihilation and creation of species. During the episodes of devastation all forms of life were extinguished; during the intervening times the world was replenished with new species.

Mid portion of the Zermatt Glacier, Switzerland. Linear ridges of rocks atop the glacier (moraines) are in process of slow transport down-valley toward the observer. Smooth rock surfaces flanking the cascading stream on either side were abraded by the glacier when it was much larger and thicker. (From Agassiz's Untersuchungen über die Gletscher, *1841*)

The Ice Age, Agassiz proposed, was brought on by the last eradication of life from the earth. As the collective body-heat of the dead organisms was dissipated, the temperature of the earth's surface dropped to freezing. Death enveloped all nature in a shroud of ice, the carcasses of the Siberian mammoths were frozen, and glaciers gathered over all of North America and all of Europe to the shores of the Mediterranean (*14*). The ice covered the summits of the Jura Mountains, and spread over the site of the Alps, which according to Agassiz were not in existence at the onset of glaciation.

Then, "as a result of the greatest catastrophe which has ever modified the face of the earth," the Alps were thrust up (*15*). Huge blocks of rock propelled skyward during this cataclysm landed upon the uparched ice and slid down its slopes for great distances. Thus were the erratics transported from their sources. Following a new generation of heat-generating organisms, "the first spring of the animals and plants living today," the earth warmed and the glaciers melted back to their present positions (*16*). According to Agassiz, "there is complete separation between the present creation and the preceding ones": present species are not the descendants of those whose destruction brought on the Ice Age.

Agassiz's cogent arguments for the fact of the Ice Age were obscured by his wild speculations regarding its cause. Members of the audience reacted variously with astonishment, incredulity, horror, or indifference (*17*). De Charpentier, in particular, was embarrassed to have his own sober reflections on glaciation expounded in a context of biological moonshine. Von Humboldt earnestly advised Agassiz to forget about glaciers and return to his fossil fishes.

Agassiz did indeed continue his research in paleontology, but also began intensive field investigations of glaciers and glacial geology. His work was concentrated in and around the Bernese Alps, Chamonix Valley, the Monte Rosa, and the Matterhorn. During the winter of 1839–40 he wrote his *Études sur les Glaciers*, which was published in both French and German editions the following fall.

The *Études* is not only a scientific masterpiece but also a handsome volume, illustrated by eighteen large plates showing glaciers and related phenomena. The book is dedicated to Venetz and de Charpentier, and the introduction acknowledges the author's indebtedness to these pioneers. Agassiz describes the changes by which snow converts to glacial ice, and systematically details the morphology of glaciers as related to crevasses, to ice pinnacles, and to the moraines that form along their sides, midparts, and ends. He recites evidence for the mobility of glaciers and summarizes historical accounts of glacial advances and retreats. Against this background he establishes the criteria for identifying areas glaciated in prehistoric times, as related to old lateral and end moraines, perched boulders, erratic blocks, and striated pavements. The areal distribution of these features, he concludes, proves that all the Alpine valleys were once occupied by glaciers which spread out over the plains to the north and south.

In 1840 Agassiz was less of a catastrophist than when he delivered his address at Neuchâtel. Though he persisted in his belief that entire organic populations have been alternately wiped out and recreated in the course of time, and that the elevation of the Alps was accomplished during the Ice Age, he soft-pedaled and relegated to a subordinate position his biological theory for the onset of continental glaciation. Moreover, he concluded that the distribution of old moraines shows that the melting back of the glaciers to their present positions "far from being sudden, has on the contrary been characterized by a variable number of periods of interruption" (*18*).

During the next twenty-eight years Agassiz spread the doctrine of continental glaciation to the British Isles and North America. Shortly after the publication of the *Études* he toured northern England, Scotland, and Ireland, everywhere finding the telltale evidence of massive glaciation. His frequent companion in these excursions, William Buckland, revised his opinion regarding the universal efficacy of Noah's Flood in remodeling the landscape, and became an enthusiastic supporter of the glacial hypothesis.

In 1846 Agassiz came to America to lecture at the Lowell Institute in Boston. In subsequent excursions he found compelling evidence for continental glaciation in New England, around the Great Lakes, and farther west in the United States.

Agassiz pursued glacial studies intermittently for thirty years after his lecture at Neuchâtel, but his interests in zoology persisted until the end of his life. Shortly after his arrival in the States he accepted an appointment as professor of zoology at Harvard. There he demonstrated unsuspected talents as a promoter and fundraiser. The zoological museum he established later became the famous Museum of Comparative Zoology. He died in 1873, and a granite boulder from the moraine of the Aar Glacier was brought over to mark his grave in Mount Auburn Cemetery.

CHAPTER THIRTEEN

The Great Tree of Life

> There is a grandeur in this view of life, with its several powers, having been originally breathed into a few forms or into one.
>
> CHARLES DARWIN

The idea that the myriad kinds of plants and animals now living are lineal descendants of one or a few simple forms which lived long ago is a unifying concept without which modern biology would be much the poorer. So indeed would be the sellers of paper and ink, for the theory of organic evolution has generated a mountain of literature since Charles Darwin's *Origin of Species* was published on November 24, 1859. Little wonder, considering that thoughtful adults can hardly fail to react in some manner on encountering the evolutionary idea for the first time. Some find it ennobling to all creatures, as did Darwin. Others concede distant kinship to archenemies or primitives of the human kind, but disavow blood relationship with lower forms of life, as exemplified in baboons, bedbugs, and bacteria. Creationists, of course, are universally hostile, and those earnest souls were much in the majority at the time Darwin exploded his bomb. That the *Origin of Species* is perhaps the most celebrated and at the same time most generally misunderstood biological treatise is a marvel second only to the fact that it got written at all.

Darwin had virtually no formal training in biology or geology. Mastery of these sciences came from his studies in the field, extensive reading, and close associations with gifted scientists. Even as a child, however, he displayed an interest in natural history, as a collector of shells, minerals, rocks, and birds' eggs. In 1817, when Darwin was eight years old, his mother died, and his father soon afterward placed him in a boarding school not far from the family home in Shrewsbury. There his interest in the out-of-doors did nothing to improve his knowledge of Latin, which prompted his father to prophesy in a moment of exasperation, "You care for nothing except shooting, dogs, and rat-catching, and you will be a disgrace to yourself and all your family" (*1*).

The father, Dr. Robert Waring Darwin, was a respected and prosperous physician, who wanted the son to follow him in his profession. In 1825 he sent Charles to the University of Edinburgh for medical training. But the young Darwin found the lectures "intolerably dull," nor could he bear to witness the surgery that was conducted in the demonstration theater without benefit of anesthetic. During his second year at the university he cultivated the acquaintance of students interested in natural history, and attended lectures by the arch-neptunist Jameson.

> . . . I attended Jameson's lectures on Geology and Zoology, but they were incredibly dull. The sole effect they produced on me was the determination never as long as I lived to read a book on geology or in any way to study the science. (*2*).

Thus did Darwin take his place, alongside his father, in the ranks of the false prognosticators. His undistinguished performance at Edinburgh did, though, convince father and son that medicine was not the right prescription in this case. Determined that Charles should not turn out to be the "idle sporting man" which seemed to be his destiny, Dr. Darwin recommended the clergy as a more honorable alternative. The idea of life as a country parson, with easy access to the woods and open spaces, was agreeable to young Darwin; and in the fall of 1827 he was admitted to Christ's College, Cambridge, to train for Holy Orders. At that time Darwin was not motivated to acquire professional training as a means of earning his bread: he confidently expected to live off his inheritance.

> Although . . . there were some redeeming features in my life at Cambridge, my time was sadly wasted there and worse than wasted. From my passion for shooting and for hunting and when this failed for riding across country I got into a sporting set, including some dissipated, low-minded young men. We used often to dine together in the evening . . . , and we sometimes drank too much, with jolly singing and playing at cards afterwards. I know that I ought to feel ashamed of days and eve-

nings thus spent, but as some of my friends were very pleasant, and we were all in the highest spirits, I cannot help looking back to these times with much pleasure. (*3*)

Of the redeeming features Darwin mentions, the most significant were his associations with the Reverend Professor John Stevens Henslow. Prior to Darwin's arrival at Cambridge, Henslow had served as Professor of Mineralogy; in 1827 he moved to the Chair of Botany. In partnership with Sedgwick he promoted programs of study in natural science, which in due course led to the granting of degrees in that subject. Henslow is also credited with establishing a new botanical garden and with transforming museum displays from collections of oddities to devices for public instruction in natural history (*4*).

The Henslows held weekly open houses for students interested in science. Darwin, who had become compulsively fond of collecting beetles at this point in his career, soon became a regular member of the group. A warm and enduring friendship grew out of this association; the two often took long walks together, and from this habit Darwin was sometimes called, perhaps enviously, "the man who walks with Henslow" (*5*). Darwin did not attend Sedgwick's lectures, but Henslow encouraged him to give geology a second chance. He arranged for Darwin to meet Sedgwick, and in 1831 Sedgwick obligingly took Darwin on a geological excursion through Wales.

Later that same year Henslow was asked to recommend a naturalist to accompany Robert Fitzroy, Captain of H.M.S. *Beagle*, on a voyage of exploration around the world. Henslow promptly and surprisingly recommended Darwin. In a letter to Darwin he candidly explained that Fitzroy was more interested in signing on a companion with the upbringing of a *gentleman* than in sharing his quarters with a "finished naturalist." Surely Darwin could do the necessary "collecting, observing and noting anything to be noted in Natural History" (*6*).

Darwin wished to accept the position, but his father offered strong objection. The voyage around the world, Dr. Darwin argued, was a wild and useless scheme, disreputable to the character of a clergyman, and yet another change of profession that would make settling down to a steady life all the harder. Furthermore he suspected that the job had been offered to and turned down by others better qualified: therefore there must be something wrong with it—uncomfortable accommodations, most likely. Fortunately for the advancement of science, Darwin's uncle, Josiah Wedgwood, interceded in his behalf. Dr. Darwin withdrew his objections, and when the *Beagle* sailed from Plymouth on December 27, 1831, Charles was aboard.

The expedition lasted for almost five years. At the Cape Verde Islands Darwin saw his first volcanic mountain, in Brazil his first tropical rain forest, and in Tierra del Fuego his first savages. He dug up bones of fossil sloths, mastodons, and horses in Argentina, and experienced his first earthquake in Chile before the *Beagle* sailed westward to Australia via the Galápagos Islands, Tahiti, and New Zealand. The remainder of the voyage took him to Mauritius, South Africa, St. Helena, Ascension Island, and back to Brazil, before the ship put in at Falmouth in October of 1836.

Darwin's vow never to read a book on geology was broken early on the voyage.

> I had brought with me the first volume of Lyell's Principles of Geology, which I studied attentively; and this book was of the highest service to me in many ways. The very first place which I visited, namely St. Jago in the Cape Verde Islands, showed me clearly the wonderful superiority of Lyell's manner of treating geology, compared with that of any author whose works I had with me or ever afterwards read. (7)

At the time when Darwin first read Lyell, neither the author nor the reader was an evolutionist. In the course of his long journey, however, Darwin saw things which cast doubt in his mind concerning the prevailing doctrine of the immutability of species. With regard to the geographical distribution of similar species, he encountered a paradoxical situation: similar animals exist far apart geographically, whereas nearly adjacent areas may have distinctive populations. The similarities between the rhea of South America and the ostrich of Africa exemplify the first case; and the distinctive subspecies of finches and tortoises on each of the Galápagos Islands provide a striking example of the second.

Soon after his return to England, Darwin realized that this paradox, as well as many other troublesome problems following on the assumption of immutable species, would be resolved if the animals in question have descended with modification from ancestral species. In July of 1837 Darwin began collecting information relevant to transmutation of species.

Work on this new project was necessarily intermittent. Over a period of about ten years following his return, his energies were spent mainly in writing reports of investigations conducted during the preceding five. He published journals of his research on natural history and geology, a celebrated book on the structure and distribution of coral reefs, and monographs on the geology of volcanic islands and on his geological observations in South America.

Writing was not Darwin's favorite exercise, and this work became the more onerous as his health inexplicably began to fail. Less than a year after his voyage he began to experience uncomfortable palpitations of the heart, and in October of 1837 he wrote Henslow complaining that "of late anything which flurries me completely knocks me up afterwards and brings on a violent palpitation of the heart" (*8*). The cause of this illness has been the subject of much speculation, and the issue may never be settled. One explanation proposes that Darwin had contracted Chagis' disease while in South America. This ailment, which is prevalent in Argentina, Chile, and parts of Brazil, is caused by parasitic protozoans transmitted to man by blood-sucking insects. Its symptoms are much the same as Darwin described. After mid-1842, at age thirty-three, he was no longer able to climb mountains or to take the long hikes necessary for geological studies (*9*).

Darwin's first major reports on the results of his studies during the voyage were published in 1839, the year he established residence in London. There he married his first cousin, Emma Wedgwood, and became active in the affairs of the Geological Society. The obligations combined in professional and social life soon proved too taxing for his diminished energies, however, and in the fall of 1842 the Darwins moved to Down House in a rural setting some twenty miles south of the city. There Darwin lived as a semi-invalid until his death in 1882.

Before Darwin left London he drafted a sketch of a theory for the origin of species, and two years later expanded his thoughts on the subject into a long essay. In 1856 he at last began writing a major work on the transmutation of species.

> . . . Lyell advised me to write out my views pretty fully, and I began at once to do so on a scale three to four times as extensive as that which was afterwards followed in my Origin of Species; yet it was only an abstract of the materials which I had collected, and I got through about half the work on this scale. But my plans were overthrown, for early in the summer of 1858 Mr. Wallace, who was then in the Malay Archipelago, sent

me an essay "On the tendency of varieties to depart indefinitely from the original type"; and this essay contained exactly the same theory as mine. (*10*)

Charles Darwin (Portrait from the DeGolyer Western Collection, Southern Methodist University)

Alfred Russell Wallace, fourteen years Darwin's junior, had dashed off his essay in the course of three evenings while recovering from an attack of malaria. His ideas had not materialized in a flash of delirious inspiration, however. He was an accomplished botanist and entomologist with field experience in the Amazon Basin; and at the time of writing his essay he had completed four years of work on the species problem in the East Indies.

Receipt of Wallace's essay posed a problem in professional ethics for Darwin. He was compelled to recommend publication on grounds of merit; on the other hand he hesitated to yield priority on ideas which he had twice drafted and was in process of expanding for publication. Lyell and Sir Joseph Hooker, the distinguished botanist, came to the rescue by having both the Wallace essay and a summary paper by Darwin read at the same meeting of the Linnean Society. Little time was lost in making this arrangement: Darwin had received Wallace's manuscript on June 18, 1858, the readings took place the following July 1, and the papers were published in the *Proceedings* of the Society soon afterwards. The published versions attracted little attention, though one critic noted that "all that was new in them was false, and what was true was old" (*11*).

Responding to the urgings of Lyell and Hooker that he publish a book on the transmutation of species as quickly as possible, Darwin abandoned the idea of completing the long version in progress; and over a period of a little more than thirteen months he wrote the summary of 1859 which we know as the first edition of *The Origin of Species.*

The book was an instant success: all copies of the first edition were sold or spoken for on the day of publication. Successive editions continued to appear as Darwin corrected and amplified earlier ones. The final text, issued nineteen years after the first edition was published, is a corrected version of the sixth edition of 1872. Darwin often complained that few readers really understood his theory. One source of this confusion arose from his constant amendations of the text. The sixth edition of the *Origin* is almost a third longer than the first; and almost three out of every four sentences in the first edition were revised in subsequent editions, some as many as five times (*12*). Nevertheless the thrust of his argument remained essentially the same. The argument may be summarized as follows.

Individuals belonging to existing species and varieties differ one from the other in varying degrees. Some of these individual characteristics are heritable, and so may be passed along from one generation to the next.

Domesticated plants and animals generally exhibit greater variation than is found in the wild species that most closely resemble them. This greater variability is clearly the result of controlled breeding—of selecting for propagation those individuals whose characteristics best suit man's use or fancy. The process of breeding new varieties from the most desirable stock is designated as *artificial selection.*

Among undomesticated species and varieties of plants and animals variation among individuals also obtains, in degrees proportional to the size of their populations and the expanse of the areas they inhabit.

Lacking any protection such as domestication might afford, organisms living in the natural state are engaged in a struggle for existence. This struggle necessarily follows from the fact that organisms tend to multiply in numbers greater than the environment can support. In the course of the contest, those variations which best adapt lineages to whatever environments are available to them are the ones most likely to be preserved in future generations. The process by which useful variations are preserved through inheritance is to be called *natural selection*—the survival of the fittest. By the long-continued operation of natural selection, favored varieties may evolve into new species, while those less fit may become extinct.

Whether by artificial or by natural selection, varieties are produced by the adding up of individual variations over many generations. Considering the spectacular results of artificial selection in producing new strains of domesticated plants and animals over the past few thousand years, the less surprising it is that natural selection, operating over millions of years, has produced all the varied kinds of organisms now living or preserved in the rocks as fossils.

The fact that all the ancestral forms of species now living have not been found in a fossilized state is mainly due to the fragmentary nature of the stratigraphic record. Within this imperfect record the sequence in which fossils occur is nonetheless orderly, and is supportive of the view that all present forms of life have evolved from one or several simple forms over vast periods of time.

As a notable example of the proliferation of varieties through artificial selection, Darwin cited the numerous different kinds of domestic pigeons. The pigeon is probably the first bird to be tamed by man; in any case there are records of its domestication in Egypt some 4,300 years ago. All the modern breeds—carriers, tumblers, runts, pouters, turbits, Jacobins, fantails, trumpeters, and the rest—derive from the still-extant blue-gray rock dove. As Darwin points out, "it is in human nature to value any novelty, however slight." In the case of pigeons, the breeders have chosen for mating individuals with unusual characteristics in conformation or color, to the end of accentuating these characters in successive generations. Crosses between different varieties produce fertile hybrids which revert to the appearance of rock doves, which indeed most street pigeons in cities over the world strongly resemble.

As with pigeons, so with the numerous varieties of domesticated dogs, horses, sheep, apples, gooseberries, and flowers in the florist's shop, Darwin argued. The breeder's maxim is simply breed from the best stock, "not indeed to the animal's or plant's own good, but to man's use or fancy" (*13*).

In a state of nature individual differences among members of the same species may not be so noticeable as among members of domesticated species. But the differences are there, Darwin insists. To make his point he cites studies in the variability of patterns formed by the nerves of certain insects and even in the musculature of their larvae.

This universal variability among individuals constituting the populations which are called species is a nuisance to the systematist, but at the same time a fact supportive of evolutionary theory. Taxonomists usually cope with this problem, Darwin submits, by naming one or more varieties within a species. But standards of classification are arbitrary: the varieties of one authority may be classed as species in their own right by another, or reduced to the rank of mere geographic races by yet another investigator. Gradational forms commonly link end members in an unbroken chain of varieties, the end members of which appear to be distinctly different species. The categories of species, variety, and race, Darwin concludes, are terms of convenience which often serve to mask the essential fact of variation among individuals. Within the evolutionary frame of reason, these individual differences are the first steps toward emergence of varieties, which in turn may be viewed as incipient species.

Darwin's ideas on the struggle for existence were reinforced by his reading in 1838 of Malthus' essay on the principle of population. The Malthusian doctrine states in essence that since human populations increase geometrically, while human food supply increases arithmetically, unchecked growth in human populations will inevitably lead to death and misery for the poor.

> A struggle for existence inevitably follows from the high rate at which all organic beings tend to increase. . . . Hence, as more individuals are produced than can possibly survive, there must in every case be a struggle for existence, either one individual with another of the same species, or with the individuals of distinct species, or with the physical conditions of life. It is the doctrine of Malthus applied with manifold force to the whole animal and vegetable kingdoms; for in this case there can be no ar-

> tificial increase of food, and no prudential restraint from marriage. Although some species may now be increasing . . . in numbers, all cannot do so, for the world would not hold them. (*14*)

For plants which annually produce a multitude of seeds, the implications of the Malthusian doctrine are self-evident; but even for slow breeders, such as man and the elephant, unchecked increase in progeny would produce unbelievably large populations over short periods of time. Darwin cites calculations which show that "if an annual plant produced only two seeds, and their seedlings next year produced two, and so on, then in twenty years there should be a million plants" (*15*). By the same line of reasoning, a population of some 19,000,000 elephants might descend from the mating of an orginal pair over a period of about 750 years.

The fact that there is still standing room for animals and growing space for plants results from the high mortality of the individuals in competition with one another. What we may offhand perceive as a peaceful pastoral scene will on closer examination reveal itself as a battleground.

> We behold the face of nature bright with gladness, we often see superabundance of food; we do not see or we forget, that the birds which are idly singing round us mostly live on insects or seeds, and are thus constantly destroying life; or we forget how largely these songsters, or their eggs, or their nestlings, are destroyed by birds and beasts of prey; we do not always bear in mind, that, though food may be now superabundant, it is not so at all seasons of each recurring year. (*16*)

The soft-hearted Darwin was not cheered by this perception. For any readers who might be appalled by the carnage implicit in natural selection, he offered sympathetic words of consolation.

> When we reflect on this struggle, we may console ourselves with the full belief, that the war of nature is not incessant, that no fear is felt, that death is generally prompt, and that the vigorous, the healthy, and the happy survive and multiply. (*17*)

Darwin emphatically denied the claim that special characteristics possessed by the healthy and happy surviviors in the struggle for existence have been impressed on them by climate, food, or other external factors in the environments they inhabit. Contrariwise, variations arise from within the organisms themselves, and those which best equip the individual to live and multiply in some particular environment are necessarily the ones most likely to be passed along to the next generation. To survive, varieties and species must adapt to their external conditions, not only to those of the inorganic world, but in most instances to a host of other creatures. Darwin cites the woodpecker and the mistletoe as examples of adaptation.

> Naturalists continually refer to external conditions, such as climate, food, &c., as the only possible cause of variation . . . but it is preposterous to attribute to mere external conditions, the structure, for instance, of the woodpecker, with its feet, tail, beak, and tongue, so admirably adapted to catch insects under the bark of trees. In the case of the mistletoe, which derives its nourishment from certain trees, which has seeds that must be transported by certain birds, and which has flowers with separate sexes absolutely requiring the agency of certain insects to bring pollen from one flower to another, it is equally preposterous to account for the structure of this parasite, with its relationship to several distinct organic beings, by the effects of external conditions, or of habit, or of the volition of the plant itself. (*18*)

Darwin recognized that natural selection is a viable concept if and only if there has been enough time for its operation. At one point he declares that "he who can read Sir Charles Lyell's grand work on the Principles of Geology, which the future historian will recognize as having produced a revolution in natural science, and yet does not admit how vast have been the past periods of time, may at once close this volume" (*19*). Naturalists may be forgiven for believing that species are immutable creations, he concedes, as long as they labored under the misconception that the earth is only a few thousand years old.

Darwin also borrowed from Lyell the idea that great changes occur in nature by the summation of small changes over protracted periods of time. The gradual step-by-step modification of organisms may at first glance seem insignificant, just as each step in the lowering of a stream bed. Yet the end results of a series of such changes may be spectacular—a new species or genus in the one case, a great canyon in the other.

> Natural selection acts only by the preservation and accumulation of small inherited modifications, each profitable to the preserved being; and as

> modern geology has almost banished such views as the excavation of a great valley by a single diluvial wave, so will natural selection banish the belief of the continued creation of new organic beings, or of any great and sudden modification in their structure. (*20*)

As evidence for the incomprehensively vast duration of geologic time, Darwin reiterates the arguments previously presented by Lyell. The more than thirteen miles in thickness of sedimentary strata preserved in the British Isles are sufficiently impressive in this connection, he submits, considering not only that this enormous mass accumulated layer by layer, but also that there are so many gaps in the record, along the numerous unconformities between successive rock units. The record of the rocks is episodic, not continuous. So with the record of life as preserved by fossils—at best "a poor collection made at hazard and at rare intervals." Therefore:

> . . . we have no right to expect to find, in our geological formations, an infinite number of those fine transitional forms which, on our theory, have connected all the past and present species of the same group into one long and branching chain of life. We ought only to look for a few links, and such assuredly we do find—some more distantly, some more closely related to each other; and these links, be them ever so close, if found in different stages of the same formation, would, by many palaeontologists be ranked as distinct species. (*21*).

Darwin regarded natural selection as a process operating in accordance with natural laws, but allowed that life may have been "originally breathed by the Creator into a few forms or into one." Even so, natural laws are human formulations—descriptions of "the sequence of events as ascertained by us" (*22*). Given a beginning, all life has descended with modification from a few simple originals. As the face of the earth has changed, attending advances and retreats of the sea, elevation and wearing down of the continents, and episodes of glaciation and volcanism, organisms have struggled in adapting to changing physical environments, and at the same time have competed with each other for food and space. In the struggle for life there have been more losers than winners; for the usual outcome of evolution by natural selection has been the extinction of species, as the record of fossils amply proves.

According to this view of life all organisms that live or have lived are kin. This idea is hard to swallow, and most of us have been taught to forgo the effort. To Darwin, however, the concept of universal kinship is not denigrating to humans but ennobling to all forms of life. In a memorable passage he compares the evolution of life to the growth of a great tree.

> The affinities of all the beings of the same class have sometimes been represented by a great tree. I believe this simile largely speaks the truth. The green and budding twigs may represent existing species; and those produced during former years may represent the long succession of extinct species. At each period of growth all the growing twigs have tried to branch out on all sides, and to overtop and kill the surrounding twigs and branches, in the same manner as species and groups of species have at all times overmastered other species in the great battle for life. The limbs divided into great branches, and these into lesser and lesser branches, were themselves once, when the tree was young, budding twigs; and this connection of the former and present buds by ramifying branches may well represent the classification of all living and extinct species in groups subordinate to groups. Of the many twigs which flourished when the tree was a mere bush, only two or three, now grown into great branches, yet survive and bear the other branches; so with the species which lived during long-past geological periods, very few have left living and modified descendants. From the first growth of the tree, many a limb and branch has decayed and dropped off; and these fallen branches of various sizes may represent those whole orders, families, and genera which have now no living representatives, and which are known to us only in a fossil state. . . . As buds give rise by growth to fresh buds, and these, if vigorous, branch out and overtop on all sides many a feebler branch, so by generation I believe it has been with the Great Tree of Life, which fills with its dead and broken branches the crust of the earth, and covers the surface with its ever-branching and beautiful ramifications. (*23*)

The point to be emphasized here is that this great tree is rooted in the expanded concept of geologic time as envisioned by Hutton and Lyell.

CHAPTER FOURTEEN

Kelvin

Whenever you can, count.
SIR FRANCIS GALTON

In 1852, seven years before *Origin of Species* was published, William Thomson, Professor of Natural Philosophy at the University of Glasgow, read a brief paper entitled *On the universal tendency in nature to the dissipation of mechanical energy*. The burden of this report is that while mechanical energy can neither be created nor annihilated through natural processes, it is constantly being transformed. And in every change of energy from one form to another a portion of the original amount is converted to heat and dissipated. Thus there can be no perpetual motion machine such as a self-winding mechanical clock: for the heat generated by friction between the moving parts constantly escapes from the system and diminishes the store of potential and kinetic energy available to swing the pendulum or otherwise move the hands. This principle of the dissipation of energy is the basis for the second law of thermodynamics, of which Thomson was co-discoverer.

At the time he presented this paper, Thomson was aware that this new principle held important implications for the history of the earth. For if the earth functions as a heat machine, as Hutton and his followers held, then it must constantly be losing energy, as heat is conducted from the interior through the crust and thence dissipated through the atmosphere. The sun too is spending its radiant energy at an enormous rate, and so, Thomson reasoned, must have been hotter in the geologic past than now, as it must surely grow cooler in the future. Extrapolating from the present backward and forward in time, Thomson envisioned a fiery beginning for the earth, and an icy end to the earth as an abode for life. Or as he put it:

> Within a finite period of time the earth must have been, and within a finite period of time to come the earth must again be, unfit for the habitation of man as at present constituted, unless operations have been, or are to be performed, which are impossible under the laws to which the known operations going on at present in the material world are subject. (*1*)

For the next forty-seven years this passage served as a text from which Thomson developed cogent arguments favoring the proposition that geologic time, far from being virtually limitless, is finite and much shorter in duration than Lyell, Darwin, and their followers wished to concede (*2*).

Thomson was born in 1824 in Belfast to parents of Scottish origin. His father, Professor James Thomson, taught mathematics at the Royal Belfast Academical Institution. In 1832 the family relocated in Glasgow, following the professor's appointment to the Chair of Mathematics at the university there. The father, an able mathematician and author of textbooks on the subject, was also a devoted tutor to his children. William in return proved a precocious student. He was admitted to the University of Glasgow at the age of ten years and three months. In the course of six sessions of study there he won prizes in classics and in mathematics. During two of the summers between sessions Professor Thomson took his family to the Continent in order that the children might learn French and German. Before William was seventeen he had mastered Fourier's *Théorie analytique de la chaleur* and La Place's *Méchanique céleste*, and had published his first paper in mathematics (*3*).

In 1841 William left Glasgow without taking a degree and enrolled at St. Peter's College, Cambridge. At that time the undergraduates there were informally classified in two groups: the "rowers," those lighthearted ones who spent much time sculling on the Cam, and the "readers," who spent most time with their studies. Young Thomson won prizes as a member of both sets; and although studies commanded highest priority he found time also to cultivate his fondness for music. One of the founders of the Cambridge Musical Society, he later became its president and played second horn in the orchestra.

Relieved of his duties as tutor, Professor Thomson now assumed the role of career-counselor to his son. William Meikelhorn, Professor of Natural Philosophy at the University of Glasgow, had been in ill health since 1838. His deterioration was not linear with time, but cyclical, rallies alternating with sinking spells. As Professor Thomson observed his colleague's struggle for life, the idea dawned in his mind that Meikelhorn might endure long enough for William to qualify as his successor. This idea soon became an obsession with the father and an agreeable prospect to the son.

Thereafter Professor Thomson's letters to William were filled with canny advice on how to train at an English institution for a professorship in Scotland (*4*). Be thrifty, he urged: use "all economy consistent with comfort and respectability . . . do not spend a sixpence unnessarily." Avoid arguments about politics and religion. Take care not only to *do* what is right, but also always *appear* to do so. More specifically, "avoid boating parties of any degree of disorderly character"—gossip travels fast and far. Improve elocution, and skill at writing for a popular audience. Continue to cultivate mathematics, but avoid the image of being simply an "expert x + y man"; show that you can also do things with the hands, such as manipulating glassware in the laboratory.

After receiving his bachelor's degree in 1845, William went to the University of Paris to bone up on experimental science. During a sojourn of less than five months he attended lectures in experimental physics and became interested in the application of mathematics to the study of electricity. Later that same year he returned to Cambridge, where he was appointed Fellow and Mathematical Lecturer of St. Peter's College.

Professor Meikelhorn finally died in 1846, and William learned of that event on May 9. Before the end of the month he filled formal application for the position thus vacated. Now working as campaign manager, Professor Thomson immediately set about pulling strings and instructing his son in the art of eliciting the right kinds of recommendations from the right people. "You must do everything that honor and propriety will allow" to get the job, he advised. Giving a hint to writers of testimonials that the position is "half-mathematical and half-popular" might guide their pens on the right tracks (*5*).

Father and son were successful. The testimonials to William's high credentials carried thirty-eight signatures, including those of all the Resident Fellows at St. Peter's College (*6*). As the letters arrived

Professor Thomson sent advance copies to influential friends among the electors. When the originals were assembled as a packet for formal presentation to the faculty, the applicant appended a list of the twenty-six scientific and mathematical papers he had already published. In September of 1846 he was named Professor of Natural Philosophy, at age twenty-two. One who saw the senior Thomson emerge from the meeting of electors observed: "A face more expressive of delight was never witnessed. The emotion was so marked and so strong that I only fear it may have done him injury" (7).

The appointment carried a proviso, however; within a month the young professor must submit a suitable essay, written in Latin, on an assigned subject. The subject selected—no doubt suggested by one of the Thomsons—was the movement of heat through the body of the earth. The text, with corrections in grammar provided by the father, was presented to the faculty at the time appointed. Immediately following the meeting, William burned the essay, though some of the ideas developed in it were the basis for his subsequent challenge on the issue of geologic time.

Laying aside all considerations of academic intrigue, there can be no doubt that the electors chose the right man. During the fifty-three years of his tenure as professor at Glasgow, Thomson acquired an international reputation as a physicist and engineer. He was instrumental in the successful completion of the transatlantic cable, an accomplishment for which he was knighted at Windsor Castle by Queen Victoria in 1866. Thereafter he became a well-paid consultant in similar projects sponsored by various nations, notably the laying of the French Atlantic cable from Brest to St. Pierre in 1869. By that time his income from fees and patents on electrical devices, such as the mirror galvanometer, had made him a man of wealth. His generosity increased with his means, and he became a patron of his university by providing funds for scholarships in experimental science.

Following the death of his wife in 1870, Thomson purchased a yacht, became an expert navigator, and invented an improved sounding device that replaced soggy hempen rope with piano wire. In the course of a voyage to Madeira on a cable-laying project he met the lady who was to become his second wife. The year following their marriage in 1874 Sir William built "Netherhall," a seaside mansion at Largs. By 1880 he had taken out seventeen patents, including an

William Thomson

improved nautical compass and a machine to take soundings from a moving ship.

Meanwhile he continued to publish at a prodigious rate. When he was elected president of the Royal Society in 1890, the number of his publications in mathematics, science, and engineering had already passed the 500 mark, and he had received honorary doctoral degrees from ten institutions in five countries (*8*).

On New Year's Day of 1892, Queen Victoria made Thomson a peer of the realm, and he became Baron Kelvin of Largs—Kelvin from the name of a stream that flows near the campus of Glasgow University. His arms carry the figures of two men: one a robed student holding a voltmeter, the other a sailor with a sounding machine. The motto: *honesty without fear* (*9*). To avoid confusing William Thomson with his father (or his brother James, who joined the faculty at Glasgow as Professor of Engineering in 1873), we shall refer to him hereafter as Kelvin.

Kelvin's numerical estimates for the duration of geological time were based on his calculations of the rate at which energy at present residing in the solar system is being dissipated as heat. This dissipation is most obvious in the case of the burning sun. Insensibly, but unremittingly, the internal heat of the earth also radiates into space, as does the frictional heat generated by the braking effect of the ocean tides upon the spin of the earth. On these considerations Kelvin could not escape the conclusions that in the geological past the surfaces of the sun and earth were hotter than now, and the rotation of the earth on its axis faster.

With regard to the original source of the sun's heat, Kelvin was uncertain. He preferred the view, attributed to Hermann von Helmholtz, that our star originally acquired its heat by the collision of smaller masses, but at present is simply an incandescent mass that is cooling (*10*). To give some idea of the magnitude of the energy thus spent, Kelvin calculated that in only 81 days the heat emitted by the sun is equivalent to the energy by the motion of the earth in her orbit of 365 days (*11*). Granted that the sun's heat originated by collision of smaller bodies, he thought it highly unlikely that the heat currently dissipated is being compensated by continued infall of meteorites.

> Now, if the sun is not created a miraculous body, to shine on and give out heat forever, we must suppose it to be a body subject to the laws of matter (I do not say there may not be laws which we have not discovered) but, at all events, not violating any laws we have discovered or believe we have discovered. We should deal with the sun as we should with any large mass of molten iron, or silicon, or sodium. (*12*)

Kelvin's calculations on the length of time that the sun has illuminated the earth yielded different results in the course of his long career (*13*). In 1855, and again in 1871, he set 100,000,000 years as a reasonable limit, and was almost certain that the figure could not be raised to 500,000,000 (*14*). Whether the heat from the sun, in its early stages of luminescence, was sufficient to sustain life on the earth was yet another question, which he addressed in 1899. He reasoned then that in the course of the sun's growth by successive collisions of its component parts, the temperature would rise to a maximum and then decline toward the present after the process of accretion had ceased. On that assumption his calculations indicated that the sun was warm enough to support some form of animal and vegetable life only as far back in time as 20 or 25 million years (*15*).

Kelvin's second line of approach to determining the length of geologic time was based upon the outward flow of heat from the solid earth. His calculations in this instance again were based both on observations and on theory. That the temperature of the outermost parts of the crust increases with depth had long been known. Gottfried Wilhelm Leibnitz had observed as much when he served as engineer in mines of the Harz Mountains beginning in 1680. His observations there and elsewhere led him to theorize that the earth is cooling from an originally molten state. Kelvin subscribed to this theory, as Buffon had long before, and set to work calculating how long a molten earth would require to crust over and cool to the levels of temperature measured in mines and wells.

He was well aware that his calculations would only provide orders of magnitude, because of uncertainties inherent in his assumptions, combined with scarcity of reliable information on the physical properties of rocks forming the earth's crust. At the time he read his landmark paper *On the secular cooling of the earth* in 1862 he candidly admitted that some of the earth's internal heat might be generated by ongoing chemical reactions, and thus might not simply be residual from an original melt. He recognized also that the rate of increase in temperature with depth varies widely from place to place, ranging between 1/110° F and 1/15° F per foot of descent, according to figures at his disposal. Finally, he acknowledged that the melting temperatures of crustal rocks had not been established with any great degree of reliability.

Granting all these sources of error, Kelvin put his mathematics and physics to work, on the assumptions that, on an average, temperature increases with depth by 1° F for every 50 feet of descent, and that the melting temperature for crustal materials is between 7,000° F and 10,000° F. On the basis of the higher of these two figures, he calculated that consolidation of the crust may have taken place about 200 million years ago. Feeding the lower figure into his formulae gave a figure of 98 million years. Making further allowance for prevailing uncertainties regarding the effect of change in temperature on the physical properties of rock, and for the wide variations in observed increase of temperature with depth, he obtained what he considered outside limits for the time past since consolidation. If consolidation had taken place less than 20,000,000 years ago, he concluded, we should have more heat underground than actually we have; if more than 400,000,000 years ago we should not have as much heat as indicated by the smallest observed increment of increase with depth (*16*).

As Kelvin intermittently returned to this problem in geophysics, he came to prefer the smaller of the two figures just cited. In his address *On geological time*, delivered in 1868, he offered an estimate of the time elapsed since the beginnings of life on the planet, as follows.

> . . . when we consider underground temperature we find ourselves driven to the conclusion . . . that the existing state of things on earth, life on the earth, all geologic history showing continuity of life, must be limited within some such period of past time as one hundred million years. (*17*)

Twenty-eight years later Kelvin further reduced this estimate, on the basis of new experimental evidence relating to the physical properties of rocks at high temperature. Originally he had assumed that common types of igneous rock, such as basalt, must be heated to at least 7,000° F (=3,871° C) before they would melt. But Clarence King of the U. S. Geological Survey had announced in 1893 that "a typical basalt of very primitive character" liquified when heated only to 1,200° C. Using this new figure in his equations reduced Kelvin's figure for the consolidation of the earth from 100,000,000 to 10,000,000 years. In 1864 Kelvin had set a lower limit at 20 million, and an upper at 400 million. He now greatly reduced the upper limit and concluded that the time of consolidation was "more than 20 and less than 40 million years ago, and probably much nearer 20 than 40." King had come out with 24 million, and Kelvin observed that he was not led to differ much from that estimate (*18*).

Kelvin's calculations of the length of geological time based on the slowing in the earth's rotation due to tidal friction yielded only outside figures for the date of consolidation. The fact that the tides produced by gravitational attraction between earth and moon exert a braking action against the earth's rotation was known before the beginning of the nineteenth century, but Kelvin was the first to show the significance of this phenomenon in reconstructions of the earth's history. In his address on geological time delivered in 1868, he pointed out that as the tidal bulges move across the oceans heat must be generated, notably in those areas of shallow water where the tides rub directly against the bottom. The effect is exactly the same as the working of a friction band against a rotating wheel: the earth is insensibly rotating slower on its axis, and the days are thus gradually growing longer. Using the figures then available, he calculated that if this slowing has averaged around 22 seconds per century, then a billion years ago the earth must have been spinning so fast that if it were then solid any unattached bodies located around the equator would have been hurled into space. Kelvin considered this process of disintegration highly improbable; more likely the earth was still molten not many millions of years ago, he concluded (*19*).

When Kelvin returned to this subject in 1897, he presented new figures as limitations on the time that the earth has existed as a solid body. Again using 22 seconds per century as the rate of retardation in spin, he calculated that 7.2 billion years ago the centrifugal force at the equator would have been four times its present amount. If a solid crust were then present, it would have been deformed into a great bulging continental ring around the equator. The fact that the continents are not so arranged suggested to him that if the earth were actually in existence so long ago it must then have been in a fluid condition. Making allowance for all uncertainties, he stated, "we may safely conclude that the earth was certainly not solid 5,000 million years ago and probably was not solid 1,000 million years ago" (*20*).

Only in his later writings did Kelvin formulate a coherent theory for the physical history of the earth. He considered it almost certain that the earth was built up of meteorites falling together. The heat generated by these collisions vaporized the meteorites to form a gaseous nebula. As the nebula cooled it shrank to form a globe of lava and liquid metals, which then began to solidify from the center outward. Immediately before consolidation of the surface, the interior was solid, except for comparatively small cells of molten material lodged between slabs of collapsed crust, and possibly for a still-molten core. When consolidation proceeded to the surface a crust of primeval granite was formed. As cooling continued, the crust was broken by gigantic shrinkage cracks from time to time, and through these the residual liquid issued as flows of lava. With cooling of the atmosphere below a critical point, rain fell to create oceans, lakes, and rivers. Now that surficial water was available, low forms of vegetable life established themselves, and in due course enriched the atmosphere with the oxygen so necessary for higher forms of life.

> If the consolidation of the earth was finished 20 or 25 million years ago, the sun was probably ready—though probably not then quite so warm as at present, yet warm enough to support some kind of vegetable and animal life on the earth. (*21*)

If the earth were once a white-hot mass, how ever did it become the abode of life? Kelvin boldly addressed this question in a presidential speech before the British Association in 1871 (*22*). He was certain on only one point: life proceeds from life and nothing but life. Life may have begun on our planet when, as Darwin had written, "the Creator breathed life into a few forms or into one." Or, perhaps, seeds and even small insects may have travelled to the earth on meteorites, after which evolution ran its course, he suggested. This second hypothesis, which was promptly branded "Kelvin's star-germ theory," was a source of dismay for some and amusement for others. Certain members of the orthodox clergy were astonished to hear Kelvin not only seem to declare himself an evolutionist, but also to set the date of creation millions of years backward in time, and not here either but on some other planet. Kelvin had evoked the image of "moss-grown fragments from the ruins of another world." Some members of his audience who had watched meteors blaze across the sky wondered how the moss escaped the holocaust. T. H. Huxley commented on the improbability that meteorites could have transported to the earth the fertile eggs of elephants or crocodiles, or for that matter the eggs of shellfish or corals. Then there was the question as to how man arrived, neatly posed by Tom Taylor in verse entitled "The Truth after Thomson."

> But say, whence in those meteors life began,
> From whose collision came the germs of man?
> Still hangs the veil across the searcher's track,
> We have but thrust the myst'ry one stage back. (*23*)

Kelvin's ideas on the origin of life were often misunderstood. He was no evolutionist in the Darwinian sense. Specifically he objected to the idea that organisms have evolved as a matter of blind chance, through the often bloody workings of natural selection. If evolution there has been, he asserted, the hand of the ever-acting Creator and Ruler must have guided the course, for we have only to look about us to see evidence of intelligent and benevolent design. As for the star-germ theory, Kelvin denied any implication of atheism, since creation of life is not the less an act of creation if conducted in some distant part of the universe rather than on earth (*24*). To the end of his days life remained a mystery and a miracle.

> Mathematics and dynamics fail us when we contemplate the earth, fitted for life but lifeless, and try to imagine the commencement of life upon it. This certainly did not take place by any action of chemistry, or electricity, or crystalline grouping of molecules under the influence of force, or by any possible kind of fortuitous concourse of atoms. We must pause face to face with the mystery and miracle of the creation of living creatures. (*25*)

The above quotation is from an address to members of the Victoria Institute, a society devoted to the cause of reconciling the findings of science with the revelations of religion. In this and similar passages Kelvin reveals a steadfast belief in divine superintendence over the workings of nature. Yet he was not entirely orthodox. Though he regularly attended the ceremonies of the church, he disliked ritual, and once observed that the High Church is "high" only in the sense that spoilt meat is so.

Kelvin thought that question marks should be placed by all dates published in the Bible except those verified by independent historical evidence. Though he could not conceive that life originated from inorganic matter, he allowed that if a naturalistic solution to the origin of life could be found then an "abnormal act of Creative Power" should not be invoked. The motto on his baronial coat of arms he did not take lightly (*26*).

Despite these eccentricities in religious matters, Kelvin was installed as Chancellor of the University of Glasgow late in 1904, and following his death three years later was buried in Westminster Abbey, next to the grave of Sir Isaac Newton.

Kelvin survived certain of his cherished theories about the earth, but he left an indelible mark on geologic thought. After Kelvin, no serious scientist entertained the idea that the earth is eternal, or throughout its long but finite period of existence has been possessed of a constant and inexhaustible store of energy.

CHAPTER FIFTEEN

A Numbers Game

> It is often said that figures can be made to prove anything; and certain it is that a series of arithmetical operations does sometimes serve as introductions to very strange conclusions. The fault, of course, is not in the tool, but in the hand that uses it.
>
> ALFRED HARKER

Kelvin's earlier pronouncements on the duration of geologic time were generally ignored by geologists and biologists. But any complacency residing in the camps of Lyell and Darwin was shattered in 1865 when Kelvin attacked uniformitarianism—the doctrine on which Lyell had based his history of the earth, from which Darwin in turn had borrowed the idea of time unlimited. Thereafter, as Kelvin successively reduced his estimates of time, from several hundreds to a few tens of millions of years, the reactions from the community of scientists ranged from meek accommodation to resolute opposition. Alternate schemes of measuring time past were devised, and the figures resulting from these exercises in arithmetic ranged between 10 million and 15 trillion years. Those longer estimates within this spectrum attracted little support; and throughout the last three decades of the nineteenth century Kelvin's calculations exerted a restraining influence on the amounts of time claimed for the evolution of the earth and the life upon it. Before the end of the century, however, the physical assumptions on which he based his calculations were being questioned. Confused by rival claims and counterclaims on time, some geologists simply withdrew from the numbers game, and concluded that though the earth must be very old its age cannot be measured in years (*1*).

Kelvin's attack on uniformitarian geology was concentrated on only one of the two propositions explicitly identified by Lyell as the essential parts of that doctrine. He was agreeable to the *method* of attempting reconstruction of the earth's ancient history in terms of change brought about by agencies and forces currently in action. He was opposed to the *hypothesis* that those agencies and forces have always operated at the same levels of energy as displayed today. His thermodynamic principle demanded that if the earth be a heat machine it must be running out of energy.

> It would be a very wonderful, but not an absolutely incredible result, that volcanic action has never been more violent on the whole than during the last two or three centuries; but it is as certain that there is now less volcanic energy in the whole earth than there was a thousand years ago, as it is that there is less gunpowder in a "Monitor" after she has been seen to discharge shot and shell, whether at a nearly equable rate or not, for five hours without receiving fresh supplies, than there was at the beginning of the action. Yet the truth has been ignored or denied by many of the leading geologists of the present day . . . . (*2*)

Kelvin's address on geological time delivered in 1868 opened with the provocative declaration that "a great reform in geological speculation seems now to have become necessary." The text of this sermon to geologists was a long passage from Playfair's *Illustrations of the Huttonian Theory* in which Playfair had elaborated on Hutton's famous phrase, "no vestige of a beginning, no prospect of an end," and which ended with the following profession of faith.

> The Author of nature has not given laws to the universe, which, like the institutions of men, carry in themselves the elements of their own destruction. He has not permitted in His works any symptoms of infancy, or of old age, or any sign by which we may estimate either their future or their past duration. He may put an end, as He, no doubt, gave a beginning to the present system at some determinate time; but we may safely conclude that this great *catastrophe* will not be brought about by any of the laws now existing, and that it is not indicated by anything which we perceive. (*3*)

Kelvin flatly denounced these statements as unacceptable, on both philosophical and scientific grounds. He then proceeded through a recitation of his arguments relating to the dissipation of solar energy, retardation of the earth's spin by tidal friction, and to the internal heat of the earth, to the conclusion that there can be no uniformity in the sense of a steady-state earth. "The earth," he declared, "is filled with evidences that it has not been going on forever in the present state, and that there is a progress of events towards a state infinitely different from the present" (*4*). On the basis of underground temperatures, he concluded, "all geological history showing continuity of life must be limited within some such period of past time as one hundred million years" (*5*).

In the year following Kelvin's lecture on time, Thomas Henry Huxley responded with a sharp rebuttal in his presidential address before the Geological Society of London. To begin with, he asserted Kelvin was mistaken in assuming that uniformitarianism is the universally accepted geologic doctrine. In fact, three more or less contradictory systems of geologic thought stand side by side in Britain; namely, uniformitarianism, catastrophism, and evolutionism. The uniformitarians advocate a practically unlimited bank of time, while the catastrophists insist upon an almost unlimited bank of force. Evolutionism, which is destined to swallow up the other two, is not antagonistic to either on all grounds, he asserted. From catastrophism, the evolutionists borrow the "idea of the development of the earth from a state in which its form, and the forces which it exerted, were very different from those we now know" (*6*). From uniformitarianism the evolutionists appropriate two concepts: the idea that a succession of small changes may accomplish major alterations over long spans of time, and the maxim that we must "exhaust known causes before flying to the unknown" (*7*). Thus evolutionism preserves the better parts of the old antagonistic doctrines, eliminates the objectionable claims of each, and applies the same fundamental way of thinking to the living and the non-living world.

Huxley went on to accuse Kelvin of kicking a dead horse. The uniformitarianism he had attacked was the outmoded and radical doctrine espoused by Hutton and Playfair. Few uniformitarians of the 1860's, Huxley claimed, would deny that the spin of the earth may be diminishing, that the sun may be waxing dim, and that the earth is cooling. If so, these matters seem to have made no practical difference during the period for which a record is preserved in stratified rocks.

As to Kelvin's proposal that all geologic history showing continuity of life must be limited to some such period of time as 100 million years, Huxley found this pretty vague. May there have been two, three, or four hundred million years? But in any case, who said that a hundred million years *might not* be sufficient for geology? Allowing

100,000 feet for the thickness of Cambrian and younger strata, and 100,000,000 years for the time of their accumulation, then the deposit would have formed at the rate of 1/1000 ft. per year, or about 1/83 of an inch. If such be the case, does it call for a revolution in thought?

Having applied his own arithmetic to the issue at hand, Huxley turned to the results of Kelvin's calculus.

> Mathematics may be compared to a mill of exquisite workmanship, which grinds you stuff of any degree of fineness; but nevertheless, what you get out depends on what you put in; and as the grandest mill in the world will not extract wheat-flour from peascod, so pages of formulae will not get a definite result out of loose data. (*8*)

Huxley was unable to assail the assumptions which were at the base of Kelvin's mathematics, but he seriously questioned the proposition that the earth is nothing but a cooling mass, quoting Kelvin's unguarded comparisons of the earth to a jar of hot water and to a globe of hot sandstone. He was certain, however, that a reform in geologic speculation was unnecessary.

> The cry for reform which has been raised without, is superfluous, inasmuch as we have long been reforming from within with all needful speed. And the critical examination of the grounds upon which the very grave charge of opposition to the principles of Natural Philosophy has been brought against us rather shows that we have exercised a wise discrimination in declining to meddle with our foundations at the bidding of the first passer-by who fancies our house is not so well built as it might be. (*9*)

Kelvin's prompt rejoinder appeared in his essay *Geological Dynamics*. He dismissed Huxley's charge that his calculations were based on "loose" data: the wide allowances he had made for any errors in assumptions or basic data had provided reasonable figures for upper and lower limits of the earth's age. He protested that he had not attacked geological speculation in general, only uniformitarianism—for which many more advocates could surely be identified than Huxley had allowed. He had no quarrel with the new doctrine of evolutionism, though it really didn't seem so novel as Huxley suggested. With regard to plutonic energy he reaffirmed that this is almost entirely thermal. Volcanos and earthquakes probably give rise to less dissipation of this energy than "the continual silent action of the conduction of heat outwards"—at a rate that can be measured (*10*). Otherwise this paper was a restatement of Kelvin's earlier positions.

The Kelvin-Huxley debate settled none of the issues under discussion, but had the effect of arousing public interest in the antiquity of the earth and related problems. Reviews of the debate appeared in the popular as well as the scientific press. Perhaps the most influential review, and surely the one most strongly supportive of Kelvin, was written by Peter Guthrie Tait, co-author of the massive *Treatise on Natural Philosophy*—a work in which "thinking is forbidden, only calculation is permitted," according to one critic (*11*). Tait's brilliance in mathematics was exceeded only by his arrogance. In his review he extolled the virtues of mathematics and physics in removing the uniformitarian blinders from the eyes of the geologists and setting them back on the path to truth. He could find no flaw in the reasoning behind his friend's calculations, and only one possible error in his figures. Kelvin's allowance of 100,000,000 years to the uniformitarians, he proposed, was but a manifestation of his generous nature: a closer examination of the data would show that only 10 to 15 million years, or less, can be allowed for the age of the earth.

By 1870 geologists were faced with the alternatives of accepting Kelvin's figures or providing better numerical estimates from evidence other than the thermodynamical. During the following three decades their schemes for calibrating the geologic time scale in years were numerous, varied as to method and results, and, as it turned out, mostly unreliable.

The most obvious method for converting relative time to years derives from the assumption that the time required for accumulation of a sequence of strata varies directly with the thickness of the sequence. This granted, it is necessary to know at what average annual rate a unit of thickness—such as an inch or a foot—has formed. Knowing the total thickness and the average annual rate of accumulation, the former divided by the latter gives the time in years represented by the sequence.

John Phillips, nephew of William Smith, applied this method as early as 1860. He assumed that sediment derived from erosion of the lands is at present accumulating over an equivalent area of the sea floor at the average rate of one foot in 1,332 years. The latter figure was taken from current estimates of the rate at which the basin of the Ganges River is being lowered annually by erosion. Taking the maximum thickness of the fossiliferous systems constituting the geologic column at 72,000 feet, the time required for the column to form would then be 72,000 × 1,332 = 95,904,000 years. Phillips himself emphasized the imprecision of this figure. The annual rate at which the Ganges Basin is yielding sediment to the sea is probably greater than the average for all the continents and islands, and if so the estimate of about 96 million is too low. On the other hand, the sediments washed off the continents may be spread over a lesser portion of the sea floor than assumed, in which case the figure is too high. Either way, he concluded, the oldest fossiliferous strata are incomprehensibly ancient (*12*).

Phillips' figure of almost 96 million years, based on geological evidence, tended to lend credence to Kelvin's pronouncement of 1868 that "all geologic history showing continuity of life must be limited in some such period of time as one hundred million years." On the other hand, Tait's proposal to reduce these estimates to ten or fifteen million was at the time unacceptable even to Kelvin. Inadvertently, the conservative and abrasive Tait had cast Kelvin in the role of a defender of the longer scale of geologic time as adduced from physical principles.

Thus C. Lloyd Morgan, writing in 1878, refers explicitly to Kelvin's liberality in *granting* to geologists about 100 million years. Morgan then went on to make his own estimates based on Phillips' method, with some refinements. He pointed out that different kinds of sediments must have accumulated at different rates in different situations. The most rapid accumulations, he believed, are taking place along deltas, in lakes and estuaries, and around live coral reefs, at annual increments of 1/8 to 1/10 of an inch per annum. The lowest annual rates of accumulation, estimated at between 1/100 and 1/256 of an inch, should occur offshore away from large deltas and in the ocean bottoms. For deposits of chalk, coal, and volcanic deposits he assumed intermediate annual rates ranging between 1/25 and 1/40 of an inch. Examining now the physical constitution of the geologic column, assumed to be 100,000 feet thick, he concluded that half accumulated at an annual rate of 1/100, a fourth at 1/50, and a fourth at 1/25 of an inch. Result: 82.5 million years for the accumulation of the entire column. In the light of all this arithmetic, Morgan wondered why geologists should balk at accepting 100 million years as the limit of geological time. As for himself, he admitted that 50 million would not make him feel very uneasy (*13*).

Alfred Russell Wallace, Darwin's junior partner in evolutionary theory, was an unlikely early convert to the Kelvin chronology. Having seen the light, however, he sought to protect the theory of natural selection from going down the drain along with Darwin's assumption of Lyellian time. Accordingly, he proposed that evolution need not have proceeded through the ages at a uniform snail's pace, but may have sped up during recurrent episodes of continental glaciation. During those hard times when the continents were being invaded by ice, life would be compressed toward the equatorial regions, the struggle for existence would intensify, and evolution would accelerate.

In 1880 Wallace offered his own calculations for the age of the earth. He accepted figures already published for the thickness of the sedimentary column (177,200 feet) and for the rate at which the continents are being denuded by erosion (one foot in 3,000 years). Unlike some of his predecessors in geological arithmetic, however, he assumed that the sedimentary waste from the continents would be spread only over that limited part of the sea floor within thirty miles of the present coastlines. If the coastlines of the world measure 100,000 miles, the area of deposition is thus $30 \times 100{,}000 = 3{,}000{,}000$ square miles, or 1/19 of the earth's total land area. Thus the area of deposition will receive sediment 19 times as fast as the land is denuded. Nineteen feet of sedimentary rock would therefore accumulate in 3,000 years, equivalent to a rate of one foot each 158 years. And $158 \times 177{,}200$ gives approximately 28 million years since the beginning of the Cambrian, leaving 72 million of Kelvin's 100 million for evolution of life during the Precambrian (*14*).

Examples of calculations similar in aim and method to those just cited could be multiplied. Between 1860 and 1900 the number of such numerical estimates increased to more than a score (*15*). Taken as a group, these show that the data from which the geologists worked were no less "loose" than those of Kelvin and Tait. Given the prevailing uncertainties regarding the maximum thickness of marine sediments preserved on the continents, the rate at which agencies of erosion transport sediment to the sea, and the area of the sea floor over which these sediments are spread, one could arrive at almost any preconceived magnitude of time. Not coincidentally, most of the figures fell within the limits prescribed in Kelvin's earlier calculations based on dissipation of the earth's heat. In this context, the proposals of the American geo-archaeologist, William J. McGee, provide a notable exception.

In 1893 McGee called attention to the fact that figures for the maximum thicknesses of the geological systems had not taken into account the stratified rocks of Precambrian age such as those exposed around Lake Superior. Adding these to the column increases the total thickness to about fifty miles. At the rate the Mississippi is pouring sediment into the Gulf of Mexico, a mile in thickness of new strata should accumulate in about 30,000,000 years. The fifty miles of the column would then account for 50 × 30,000,000, or 1,500,000,000 years. To this figure must be added the time required for the earth to cool and crust over, prior to the time that water could begin its work of sedimentary transport.

McGee was not prepared to place much faith in this simple arithmetic. He argued that the farther back in time we attempt to apply numbers for duration, the more uncertain our results. Therefore it is necessary to apply a "safety factor" to all estimates, and this factor will increase exponentially with time. For estimates of time since the last glaciation, he applied a factor of four, which increases to 64 for the Cenozoic, 256 for the Mesozoic, and 1,024 for the Paleozoic. His table of results is reproduced here (*16*).

	Mean Estimate	*Factor of Safety*	*Minimum*	*Maximum Estimate*
Post-glacial	7,000	4	1,175	28,000
Cenozoic	90,000,000	64	1,406,000	5,760,000,000
Mesozoic	300,000,000	256	1,172,000	76,800,000,000
Paleozoic	2,400,000,000	1024	2,343,000	2,457,600,000,000
Age of the Earth	6,000,000,000	—	10,000,000	5,000,000,000,000

McGee judged his maximum figures to be as probable as his minimal, and his "mean estimates" more probable than either. Given such a wide choice, everybody could find a figure to his liking. Tait could have his 10 million as a minimal age of the earth, and any surviving disciples of de Maillet could rejoice in the five trillion allowed as a maximum.

Wide-ranging as these figures were admitted to be, McGee insisted that they were based on operations taking place in the real world and not in some imaginary planet invented by Scottish physicists. The physicists and astronomers had persistently assumed that the earth is a homogeneous body, simple in structure, which manifestly is not the case. Consequently their calculations based on these false assumptions represent "nothing more than grist ground from a mathematical mill."

> The geologic estimates concerning the age of the earth are based on real processes and actually observed conditions in such a manner as practically to eliminate inaccuracies growing out of complex and unknown factors, and are thus strictly pertinent to the case; while the non-geologic estimates are based on ideal conditions immeasurably simpler than those actually attending a planet, and thus, interesting and instructive as they are in the abstract way, have very little to do with the concrete case. (*17*)

McGee could question the quality of the product that had come out of Kelvin's mathematical mill, but it remained for John Perry to demonstrate Huxley's proposition that different kinds of grist can be ground from the same mill. Perry was a skillful mathematician and engineer who had served for a time as Kelvin's assistant and co-worker. Some of his friends had urged him to review and evaluate Kelvin's calculations, and in 1895 he did so, in a letter published in *Nature* (*18*).

Perry confessed at the outset that he disliked to consider any quantitative problem in geology. "In nearly every case," he pointed out, "the conditions given are much too vague for the matter to be in any sense satisfactory, and a geologist does not seem to mind a few million years in matters relating to time." But if some of the geologists and paleontologists are justified in wanting more than 100 million years for the age of the earth, he concluded, then there must be something wrong with Kelvin's assumptions.

The assumption that Perry first challenged was that the earth is a homogeneous mass of rock with the same capacity for conducting heat as the rocks found on its surface. If this assumption be false, then so must be Kelvin's figures for the age of the earth. Considering the wide range in physical properties of the rocks available for our inspection, combined with our ignorance of what actually constitutes the earth's interior, it would be just as reasonable, Perry submitted, to assume that the interior is made of better conducting material than that constituting the outermost skin in which the temperature-gradient is alone measured. If so, Kelvin's figures will have to be revised upward.

Perry then proceeded to examine several mathematical models for a heterogeneous earth. One of these supposes that the interior is a solid homogeneous sphere uniformly heated to 7,000° F overlain by an initially cold shell. How long, then, would be required for the temperature-gradient to attain the value used by Kelvin in his calculations, assuming that the conductivity of the inner sphere is greater than Kelvin's value for the shell? Perry showed that under these conditions Kelvin's figures for the age of the earth would have to be increased by two to six times the ratio of the internal conductivity to the conductivity of the shell.

The case for Kelvin gets worse, Perry showed, if it be imagined that the inside of the earth is not all solid, but instead "a honeycomb mass of great rigidity partly solid and partly molten." Then the effective conductivity inside would surely be greater than that of the shell. If only 10 times as great, Kelvin's age of the earth would have to be multiplied by 56.

Responses from Tait and Kelvin were printed along with Perry's letter. Tait was flip: he professed not to understand what Perry was driving at. Why drag in mathematics at all, he queried, since it's perfectly obvious that if the interior is a better conductor than the skin, Kelvin's figures must be incorrect? But Kelvin could as well have assumed that conductivity *diminishes* with rise in temperature. What then? In any case, Tait thought, the *advanced* geologists would be no more satisfied with ten billion than with a hundred million years (*19*). Kelvin was more friendly, acknowledging that perhaps he should have placed the upper limit for the earth's age at 4,000 million rather than 400 (*20*).

In 1895 Perry again returned to calculations of the earth's age, as derived from physical, geological, and paleontological information. On reviewing the evidence, he asserted that Kelvin's higher limits of a thousand million, four hundred million, and five hundred million years, as arrived at from three different approaches, very probably are all underestimated. "If palaeontologists have good reasons for demanding much greater times," he concluded, "I see nothing from the physicists' point of view which denies them four times the greatest of these estimates" (*21*).

Perry's allowance that the earth's interior might be fluid reflected the earlier views advanced by the Reverend Osmond Fisher, a prolific writer. Fisher's *Physics of the Earth's Crust*, published in 1881, is a landmark work in the emergence of geophysics as a discipline. In this book, and in articles that appeared later, he assailed Kelvin's basic assumption that the earth is solid. On physical grounds he contended that the motions displayed by the earth related to precession, nutation, and the response of the crust and waters of the ocean to the pull of the moon's gravity are equally as well explained by the assumption that the crust rests on a viscous substratum as by the assumption that the earth possesses a steel-like rigidity throughout (*22*). The Fisher model would permit internal heat-transfer by convection, and so would inflate Kelvin's figures for age in the manner indicated by Perry. Fisher also marshalled geologic evidence in support of his theory. Old shorelines appear in Scandinavia in stair-step sequences up to elevations as much as 700 feet above the present level of the sea. Until fairly recently Scandinavia was buried by glacial ice. The elevated strandlines can be explained if the weight of the glacier depressed the crust beneath, and then as the weight was removed through melting the crust bulged back to something near its original position. If so, this behavior argues for the presence of a yielding substance below the crust that can move away from beneath regions where it is overloaded from above, and then move back after the load is removed (*23*).

> With respect to the yielding of the crust, I think we cannot but lament, that mathematical physicists seem to ignore the phenomena upon which our science founds its conclusions, and, instead of seeking for an admissible hypothesis, the outcome of which, when submitted to calculation, might agree with the facts of geology, they assume one which is suited to the exigencies of some powerful method of analysis, and having obtained their result, on the strength of it bid bewildered geologists to disbelieve the evidence of their senses. Such appears to most of us the conclusion, that the earth is excessively rigid from its centre to its surface. For we know that down to quite a recent period it has yielded freely to pressure. (*24*)

As already noted, Kelvin was not moved by the arguments of Fisher, Perry, and the others who had attacked the assumptions underlying his estimates. In 1897, when he addressed the Victoria Institute, he tilted toward the findings of Clarence King and allowed only 20 to 25 million years for the earth as an abode of life. This address was reprinted in several journals, on both sides of the Atlantic, and thus provided a maximum of irritation among members of the geological fraternity, including those who had managed to adjust their chronologic sights to Kelvin's hundred million years.

Thomas Chrowder Chamberlin (Portrait reproduced from NAS Biographical Memoirs, *with the permission of the National Academy of Sciences, Washington, D.C.*)

One of those more irritated was Thomas Chrowder Chamberlin, professor of geology at the University of Chicago, who promptly responded with a lengthy critique of Kelvin's address. He began on a conciliatory note, acknowledging the great physicist's contributions in curbing "extravagant time postulates of some of the earlier geologists." Nevertheless he did not care for the "air of retrospective triumph" or the "tone of prophetic assurance" that permeated the address. Surely Kelvin had pursued one line of reasoning logically and rigorously, but "the fascinating impressiveness of rigorous mathematical analysis, with its atmosphere of precision and elegance, should not blind us to the defects of the premises that condition the whole process" (*25*).

Preliminaries tended to, Chamberlin went to work on the premises behind the mathematics. As to the assumption that the earth was initially a white-hot liquid mass, supporting evidence from geology is missing, he observed. On the contrary, the oldest rocks forming the basement complex do not remotely resemble a formerly molten envelope of the primitive earth solidified *in situ.* To go back a step farther, Chamberlin pointed out, the idea of a molten earth is based on the observation that temperatures increase with depth, combined with uncritical acceptance of the nebular hypothesis. But if the earth formed by the aggregation of meteorites, then the internal heat may have resulted from the gravitational concentration of these bodies. And if the rate of accretion was slow, then the earth may have remained solid throughout all stages of its growth. Thus the assumption that meteorites falling together would produce a gaseous or white-hot liquid is gratuitous. The outcome here would depend on the *rate* of infall, and if that rate was the same in the past as it is today, the meteorites would not liquify.

Chamberlin regarded Kelvin's assumptions of high speed of rotation of the earth during early stages of its history as an offspring of the nebular hypothesis. Surely there is no geological evidence for an equatorial landmass or for concentration of oceanic water around the poles such as must have existed if the earth has rotated so rapidly as Kelvin's extrapolations suggest. "It is not easy to see how such heterogeneity as is required for the continents and ocean basins could arise from a white-hot liquid-surfaced earth descended from a gaseous earth."

As for any calculations of the sun's age, based on present rates at which solar energy is spent, Chamberlin questioned that such can be made, considering our ignorance concerning the source of that energy. Is it possible that some of this energy is coming from within the atoms themselves?

> Is present knowledge relevant to the behavior of matter under such extraordinary conditions as obtain in the interior of the sun sufficiently exhaustive to warrant the assertion that no unrecognized sources of heat reside there? What the internal composition of the atoms may be is yet an open question. It is not improbable that they are complex organizations and the seats of enormous energies. Certainly, no careful chemist would affirm either that the atoms are really elementary or that there may not be locked up in them energies of the first order of magnitude. No cautious chemist would probably venture to assert that the component atomecules, to use a convenient phrase, may not have energies of rotation, revolution, position and be otherwise comparable in kind and proportion to those of a planetary system. Nor would he probably be

prepared to affirm or deny that the extraordinary conditions which reside in the center of the sun may not set free a portion of this energy. (*26*)

The age of the earth remained a subject of much confusion through the end of the century. In 1899 the Irish geologist, John Joly, estimated that the oceans are about 90 million years old, based on figures for the total amount of sodium in sea water and the amount annually flushed out of the lands through rivers. In his presidential address before the British Association's Section of Geology, William J. Sollas identified Kelvin as one of the great forerunners of geology, along with Steno and Hutton, for having brought to the aid of our science "the power of the higher mathematics" and for "instructing it in the teachings of modern physics." Ironically, he paired Kelvin with Darwin as the fathers of evolutional geology. "We are content," he affirmed, "to accept the infant earth . . . as a molten globe ready made, its birth from a gaseous nebula duly certified" (*27*). Sollas' own higher mathematics led him to 50 or 60 million years for the earth's age, of which 26 million were allowed for time since the Cambrian.

This confusion of numbers continued long into the twentieth century. Horace B. Woodward, writing in 1911, ended his book on the history of geology on a note of resignation. "We may safely regard the period of geological time as vast," he concluded, "but the reckoning of geological ages into any exact number of years is beyond our ken" (*28*).

CHAPTER SIXTEEN

Radiometric Dating

When a distinguished but elderly scientist states that something is possible, he is almost certainly right. When he states that something is impossible, he is very probably wrong.

A. C. CLARKE'S FIRST LAW

In 1895, the year Perry began his attack on Kelvin's calculations, Wilhelm Conrad Röntgen discovered x-rays, and thereby set in motion a train of events which revolutionized physics and medicine, and incidentally expanded concepts of geologic time into the range of thousands of millions of years.

Röntgen's discovery, for which he later received the first Nobel Prize for physics, was accidental. While experimenting with the passage of electrical currents through glass tubes filled with gases at low pressures, he noticed that a nearby specimen of a barium compound gave off light whenever the current was turned on. To the then unknown kind of energy responsible for this effect he gave the name x-rays. Röntgen further established that x-rays pass readily through many kinds of opaque materials, a fact promptly put to work in medical diagnosis.

In 1896 the French physicist Henri Becquerel found that compounds of the element uranium (which had been known since 1789) spontaneously emit energy which possesses penetrating properties similar to those exhibited by x-rays. Intrigued by this discovery, Marie Curie began studies of natural substances which exhibit the Becquerel effect. She found that pitchblende, an impure variety of the mineral uraninite, is more vigorously emissive of energy than specimens of pure uranium. On the suspicion that there must be an unknown active substance in the pitchblende, she worked in cooper-

Ernest Rutherford (Portrait courtesy of The Royal Society)

ation with Pierre Curie, her husband, to search for it. In 1898 the Curies announced the discovery of two new elements, radium and polonium. Madame Curie's term, "radioactivity," has since been universally applied to the spontaneous emission of energy and subatomic particles from substances either naturally instable or artificially made so.

Radioactivity soon developed into a new branch of physical research, largely due to the work of Ernest Rutherford, who has been called the father of the nuclear age (*1*). A New Zealander by birth, Rutherford arrived at Cambridge in 1895 for graduate studies in experimental physics at the Cavendish Laboratory. Shortly after the announcement of Röntgen's discovery later in that same year, he began experiments in passing x-rays through gases. His work over the next four years established that Becquerel rays are not simply x-rays naturally produced by the uranium or radium present in certain minerals, but rather represent a combination of a radiation designated as gamma-rays, similar in character to x-rays, together with rapidly moving streams of material particles. By passing emanations from radium through a magnetic field, he demonstrated that the particles emitted are of two kinds: one bearing a positive charge, the other a negative. The positive particles, which he called alpha-rays, later turned out to be the nuclei of helium atoms. The negatively charged particles proved to be electrons. As early as 1902, Rutherford and his associate, Frederick Soddy, framed a general theory of radioactivity. This *disintegration theory*, in their words, states that "the atoms of the radioactive bodies are unstable, and a certain fixed proportion of them become unstable every second and break up with explosive violence, accompanied in general by the expulsion of an α- or β-particle. The residue of the atom, in consequence of the loss of an α-particle, is lighter than before and becomes the atom of a new substance quite distinct in chemical and physical properties from its parent" (*2*). In the minds of some conservative scientists, the disintegration theory smacked of alchemy.

In 1903 Pierre Curie and his associate Albert Laborde announced their discovery that compounds of radium constantly give off heat. And soon afterward, Rutherford established that the amount of heat released by radioactive substances is proportional to the number of alpha particles they emit. The generation of heat by radioactivity held important implications for geology, as Rutherford made clear in an address delivered before the Royal Institution in 1904 and in a popular article published in *Harper's Magazine* the following year. To emphasize the magnitude of the thermal energy so generated, he drew a comparison between radium and coal.

> The amount of heat emitted from radium is sufficient to melt more than its weight of ice per hour. The rate of heat emission is continuous and, so far as observation has gone, does not decrease appreciably with the time. In the course of a year, one pound of radium would emit as much heat as that obtained from the combustion of one hundred pounds of the best coal, but at the end of that time the radium would apparently be unchanged and would itself give out heat at the old rate. It can be calculated with some confidence that, although the actual amount of heat per year to be derived from the radium must slowly decrease with the time, on an average it would emit heat at the above rate for about one thousand years. (*3*)

In his address Rutherford pointed out that radioactive matter has been found to be widely disseminated, not only in the rocks forming the earth's crust, but in the air within soils or in wells, and in the water of springs. Assuming then the same increase of temperature with depth and the same conductivity of rocks as used by Kelvin in his calculations, Rutherford estimated that the amount of heat conducted to the earth's surface each year and there dissipated could be supplied by radioactive matter in the extent of about five parts in ten thousand million by weight. This, he noted, is about the present concentration of radioactive matter in soil.

> Thus it does not appear improbable that the temperature gradient observed in the earth may be due to the heat liberated by the radioactive matter distributed throughout it. If this be the case, the present temperature gradient may have been sensibly constant for a long period of time, and Lord Kelvin's computation may only supply the minimum limit to the age of this planet. Thus the earth may have been at a temperature capable of supporting animal and vegetable life for a much longer time than estimated by Lord Kelvin from thermal data. Similar considerations apply to the question of the sun's heat; for the presence of radium in the sun, to the extent of about four parts in one million by weight, would of itself account for the present rate of emission of heat. The discovery of the radioactive elements, which in their disintegration liberate enormous amounts of energy, thus increases the possible limit of the duration of life on this planet, and allows the time claimed by the geologist and biologist for the process of evolution. (*4*).

Ironically, at the time this statement was made, few geologists and biologists were claiming much time beyond a hundred million years. But, soon afterward, developments from continued investigations of radioactivity raised the possibility that radioactive minerals might serve not only as sources of heat but also as timepieces to date the rocks that contained them. Given the rate, in years, at which a parent element transmutes to a daughter element as an end product, then the ratio between daughter and parent will give the age of a specimen containing the two. For the age to be reliable, the yearly rate of decay must be known, and the evidence that no significant part of either element has escaped from the specimen must be substantial. For the age to be significant in determining the date for some event in earth history, the specimen should come from rocks whose age relative to other rocks is established on independent geological evidence.

Rutherford's book, *Radio-Activity*, published in 1904, stated the principles of the subject as then understood. He offered evidence to show that the rate of decay exhibited by radioactive substances is constant and unaffected by change in temperature or other external conditions. Although the proof that the alpha particles consisted of helium atoms was to come later, he was already convinced that helium is an end product of radioactivity—in which case the relative amount of that gas trapped in radioactive minerals might yield figures for the ages of the minerals themselves. In 1905, John William Strutt (later, Fourth Baron Rayleigh) co-discoverer of the inert gas argon, obtained an age of 2,000 million years on the basis of the helium content of a specimen containing radium (*5*). That same year the American physical chemist Bertram Boltwood, noting that lead is always associated with ores of uranium, speculated that lead might be another end product in the uranium series. Using as a base the ratio of lead to uranium he calculated the ages of 43 minerals and obtained results ranging between 400 million and 2,200 million years (*6*). Neither Strutt nor Boltwood claimed precision for the rates they had assumed for the production of helium in the first instance or of lead in the second. On the other hand their data were no more "loose" than those on which Kelvin had based his calculations, and their figures indicated the wide margin by which Kelvin had underestimated the earth's age.

Kelvin was not moved. He had regarded the original announcement of Röntgen's x-rays as a hoax. Fascinated with radium, he sometimes carried about in his vest pocket a glass tube containing a specimen of the element presented to him by Pierre Curie (*7*). Yet he denied that radioactivity could be a spontaneous source of heat, whether in the earth or in the sun. Whatever thermal energy might seem to derive from radium, he contended, was actually stored there through conduction or radiation of external energy that ultimately traces back to sources in gravitation.

When Rutherford spoke before the Royal Institution in 1904, Kelvin was in the audience. Rutherford later wrote the following recollection of this encounter.

> I came into the room, which was half dark, and presently spotted Lord Kelvin in the audience and realised that I was in for trouble at the last part of my speech dealing with the age of the earth, where my views conflicted with his. To my relief, Kelvin fell fast asleep, but as I came to the important point, I saw the old bird sit up, open an eye and cock a baleful glance at me! Then a sudden inspiration came, and I said Lord Kelvin had limited the age of the earth, *provided no new source* (of energy) *was discovered*. That prophetic utterance refers to what we are now considering tonight, radium! Behold! the old boy beamed upon me. (*8*)

Announcement of the large figures for ages by Strutt and Boltwood was greeted with skepticism, mingled with relief, by most of the geologists who had wrestled with the problem of the earth's antiquity. The discrediting of the low estimates for age proposed by Kelvin and Tait was welcomed, but a conceptual leap from twenty or so millions to a thousand or so millions of years could not be taken without some careful looking. In any case, the expanded concept of time would have to jibe with geological evidence. As T. Mellard Reade put it in 1906:

> The bugbear of a narrow physical limit to geological time being got rid of, we are free to move in our own field of science. The methods of geology have this advantage over pure physics, we can more readily appeal to nature for proof or disproof. (*9*)

John Joly shared this sentiment. He appealed once again to the evidence provided by accumulation of sodium in the ocean as providing proof that Strutt's figures must be far too great. On reviewing his own figures in 1909, he increased his estimate for the age of the ocean to 110 million years. That same year Sollas raised the ante to 150 million; but in 1910 a recession set in, and other investigators using Joly's method produced figures between 74 and 80 million (*10*).

Boltwood had not tied his figures for ages of minerals to specific systems within the framework of the standard stratigraphic column. He only claimed that his dates were "not contradictory to the order of the ages attributed by geologists." In 1911 Arthur Holmes, working in Strutt's laboratory at Imperial College, was able to compare numerical ages for specimens of radioactive minerals with their relative ages as determined on geological grounds. Specimens of Carboniferous age yielded figures of 340 million years, as compared with 370 million for those of Devonian age, and 430 million for those that were of either Ordovician or Silurian age. Specimens older than the Cambrian fell in the range of 1,025 to 1,640 million years. Holmes' conclusion proved to be prophetic.

> Wherever the geologic evidence is clear, it is in agreement with that derived from lead as an index of age. Where it is obscure, as for example in connection with the pre-Cambrian rocks, to correlate which is an almost hopeless task, the evidence does not, at least, contradict the ages put forward. Indeed, it may confidently be hoped that this very method may in turn be applied to help the geologist in his most difficult task,

that of unravelling the mystery of the oldest rocks of the earth's crust; and further, it is to be hoped that by the careful study of igneous complexes, data will be collected from which it will be possible to graduate the geological column with an ever-increasingly accurate time-scale. (*11*)

Holmes was twenty-one years old when he wrote these lines, and for more than half a century afterward he did much to accomplish the objectives expressed in them (*12*).

Born at Hebburn-on-Tyne in 1890, Holmes first learned of the debate about the earth's age from Kelvin's *Popular Lectures and Addresses*, which he read in his school days. His high marks won for him a scholarship to study physics at Imperial College, where Strutt was professor. On completion of his undergraduate studies he was named Associate of the College and engaged in post-graduate work in Strutt's laboratory. Meanwhile he had become interested in geology, and so joined an exploratory expedition into Mozambique. His field work there sharpened his interest in petrology, a branch of science in which he later became expert. In 1912 he returned to London, and for the following eight years served as Demonstrator in Geology at Imperial College.

Holmes' first book, *The Age of the Earth*, appeared in 1913. In this work he reviewed, critically and evidentially, various estimates of age based on astronomical, geological, geothermal, and radiometric methods. He emphasized the discrepancies between ages based on present average rates of erosion, deposition, and sodium enrichment of sea water compared with those based on radioactivity. Something must not be uniform, he submitted. Either the rate of uranium decay is not constant, or else the average rates of change produced by geological processes at present are not reliable indicators of changes that have taken place in the past.

Given the choice, Holmes opted in favor of the radiometric ages. If laboratory experiments had demonstrated that some radioactive substances transform at constant rates, he thought it reasonable to suppose that constancy in rate of decay is a generic characteristic of all such substances. In any case, he submitted, the geologists have been incautious in assuming that *present* rates of lowering of the continents by erosion and of deposition of sediments in the oceans have prevailed always. Great linear belts of mountains, such as those constituting the Alps and their continuations into Asia, have been uplifted in comparatively recent times, geologically speaking, and so

Arthur Holmes (Portrait courtesy of Donald B. McIntyre)

the average rates of denudation and sedimentation today are in all probability higher than the average for all time since the Cambrian. Even so, a re-examination of the geologists' "hourglass" method would show that the results for the duration of geologic time are less different from the results from radiometry than has been supposed.

When Holmes was on his African expedition in 1911, he had made his own calculations on the duration of geologic time based on sedimentation. Taking Sollas' figure of approximately 50 miles as the aggregate thickness of Cambrian and younger sedimentary rocks, Holmes assumed that the sediments were deposited along the continental shores to a distance of 35 miles seaward. Taking 100,000 miles as the average shoreline of the Cambrian and subsequent periods, the total volume of the sediments equals $35 \times 50 \times 100{,}000 = 175{,}000{,}000$ cubic miles.

> The average rate of continental denudation is probably 1 foot in 5000 years, or 1 mile in 26 million years. The continental area is 56 million square miles, one-quarter of which is occupied by igneous and pre-Cambrian rocks, *i.e.* 14 million square miles. The rate of denudation of the latter is therefore 14 million cubic miles in 26 million years. The time for all the sediments to collect at that rate would be 325 million years. This figure is only an indication of the order of the time elapsed. (*13*)

At the time this was written, Holmes was estimating the duration of time elapsed since the beginning of the Cambrian at 500 million years, based on analyses of radioactive minerals. His figure of 325 million, based on geological evidence, though still short of that mark, was more than twice as large as Sollas' current estimate for the age of the ocean. Holmes listed three factors which would tend to decrease his figure and eight that would raise it; and although he made no effort to weigh these factors against each other, it seems clear that he regarded 325 as a minimal figure. Among the considerations that would increase the estimate, he cited the probabilities that the present continents are abnormally expansive, high, and rugged; that certain unconformities in the stratigraphic sequence may not be represented by sediments elsewhere, and that some sediments buried at great depths may have melted and turned into igneous rocks.

Holmes' first effort at producing a scale of geologic time calibrated in years was published in 1927. The scale was based on only 23 fixed points based on uranium-lead and uranium-helium ratios, and the figures ranged from 1,260 million years for the Precambrian to 35 million years at an horizon in the mid-Tertiary. Ten years later he presented a revised version based on 12 determinations by the lead method and 18 by the helium method. Few of these fixed points fell where the geologist would like to find them: namely, on the boundaries between the geologic systems. Approximate dates for the beginnings and terminations of geologic periods were arrived at by interpolation between the fixed points, guided by thickness, fossil content, and structural relationships of intervening strata. The results are shown in the following table, which compares Holmes' estimates of 1937 with his later revisions.

Successive changes in Holmes' figures reflect growth in knowledge of global geology combined with improved techniques of radiometric dating. In 1905 when Boltwood calculated his ages from ratios between lead and uranium, it was assumed that both these elements were internally homogeneous. Seven years later Frederick Soddy, a

Successive Time Scales developed by Arthur Holmes
(age to base in millions of years) (*14*)

	1937	*1947*	*1960*
Cenozoic	68	58–68	68–72
Cretaceous	108	127–140	130–140
Jurassic	145	152–167	175–185
Triassic	193	182–196	220–230
Permian	227	203–220	265–275
Carboniferous	275	255–275	345–355
Devonian	313	313–318	390–410
Silurian	341	350	430–450
Ordovician	392	430	485–515
Cambrian	470	510	580–620

former student of Rutherford, demonstrated that certain elements may exist in two or more forms which are indistinguishable from each other chemically but which have slightly different atomic masses. These closely allied variants he called *isotopes.* Subsequent investigations have shown that among the 106 elements now known there are more than 300 isotopes that occur naturally, including some 65 radioactive ones. In the case of lead, for example, eleven isotopes are now known, and all but one are products of radioactive decay.

Isotopes are designated according to the name or symbol of the element to which they belong, to which is affixed a number signifying the mass of elementary particles forming the nuclei of their atoms. According to modern physical theory, atoms of all elements except hydrogen contain a nucleus of positively charged particles (protons) packed together with uncharged particles (neutrons) about which negatively charged particles of much smaller mass (electrons) move in orbits. Atoms of different elements contain different numbers of protons, and the number of these gives the distinguishing atomic number for the element. Atoms of the same element contain the same number of protons, but the number of neutrons will be different for each isotope of the element. The sum of protons and neutrons gives the mass number of the isotope. In the case of the lead isotopes the mass numbers range from 203 to 214. Of these lead-204 is the only non-radiogenic variety.

These concepts were of course not available during the pioneering days of radiometric dating. Nor were the several lineages by which radioactive substances transmute to stable end-products known, nor the precise rates at which daughter substances derive from parents. For example, Joseph Barrell, Yale professor of geology, obtained dates back to 1,800 million years from minerals that contained negligible quantities of the common lead-204. But he could not know then that some of the radiogenic lead in his samples derived from the decay of thorium and not uranium (*15*).

Vast improvements in isotopic analysis and more accurate determination of decay constants attended the developments in mass spectrometry after 1927. Mass spectrometers are devices for sorting out isotopes and determining their relative abundance by passing volatilized mixtures of these through a strong magnetic field. Improved models assembled by Alfred Nier in the 1930's provided the data necessary for reliable calculations of age from isotopic analysis, and for determining the decay constant for the comparatively rare uranium-235.

Radioactivity is now understood as a phenomenon common to atoms with nuclear particles in unstable constellations. The process of decay, accompanied by emission of particles or electromagnetic energy, may proceed through a series of stages before stability is reached in an end-product. The rate of decay for a radioactive isotope is expressed in terms of its half-life, that is to say the time required for half of any given quantity of nuclei to decay. Half-lives have been found to be constant for each isotope, and may range in duration from rates too short to rates too long for accurate measurement. The radioactive series that have been of most use in geologic dating are shown in the following table.

Radioisotope	*Half-life (yrs.)*	*End product*
Uranium-238	4,500,000,000	Lead-206 and Helium-4
Uranium-235	710,000,000	Lead-207 and Helium-4
Thorium-232	15,000,000,000	Lead-208 and Helium-4
Rubidium-87	50,000,000,000	Strontium-87
Potassium-40	1,300,000,000	Argon-40 and Calcium-40
Carbon-14	5,570	Nitrogen-14

Radiometric dating was not generally accepted as a reliable method for calibrating the geologic time scale until about a quarter of a century after Becquerel's discovery. In the early 1920's conferences in Scotland and in the United States brought together experts in geology, astronomy, physics, and paleontology to discuss various approaches to determining the age of the earth. The consensus was that radiometric dating provides a reliable method for measuring geologic time in years.

In 1931 a committee for investigating the age of the earth, working under the auspices of the United States National Academy of Sciences, published a report which put an end to the abbreviated versions of geologic time developed during the late nineteenth century. The committee was interdisciplinary, its members combining competence in geology, paleontology, astronomy, and physics.

In this report Charles Schuchert wrote on the age of the earth as indicated by fossils and sedimentary rocks. Starting from scratch, he gathered figures for the maximum thickness of Cambrian and younger strata in North America and came up with a total of 259,000 feet—about the equivalent of the familiar 50 miles. Combining the maximum figures for the several systems as determined in North America and Europe, the total rises to 308,000; and Schuchert guessed that when all figures for maximum thickness are gathered from around the world the total will stand at not less than 400,000 feet. The question he addressed was whether this thickness of strata would be adequate to accommodate the 500 million years suggested by radiometric dating to have lapsed since the beginning of the Cambrian. Making allowances for gaps in the sedimentary record, as related to unconformities and breaks in deposition along bedding surfaces, his answer was affirmative. After working his way through long columns of figures for thickness, Schuchert concluded that of the 500 million years, 60 should be allotted to the Cenozoic, 120 to the Mesozoic, and 320 to the Paleozoic.

> In conclusion, the writer will admit that he is surprised over his own results, for he started with the idea that he could not find enough thickness of strata or enough breaks to meet the demands of time indicated by the radioactive minerals. He has, however, found easily enough marine strata since the beginning of Paleozoic time to call for 500 million years. In his "Historical Geology," 1924, he said . . . "that the earth since the beginning of the Archeozoic is probably at least 500 million years old." And now we are willing to admit 500 million years alone back to the beginning of Paleozoic time. One stratigrapher at least has wholly gone over into the camp of the radioactive workers! (*16*)

The workers, of course, were not radioactive, but Schuchert joined them anyhow. So did Adolph Knopf, who chaired the committee, and who, after reviewing Joly's method for measuring the age of the ocean, concluded that it was too unreliable to serve as a check on other methods (*17*). Arthur Holmes, the only committeeman who could claim charter workership, wrote most of the report. In a masterful summary on radioactivity and geological time, he reviewed the history of radiometrics, detailed the conditions to be fulfilled by any method of measuring geologic time in years, candidly described the difficulties and uncertainties inherent in radiometric dating, explained the procedures of dating minerals based on helium and lead ratios, catalogued and described the essential characteristics of 93 different kinds of natural radioactive substances, and assembled the results of radiometric dating for specimens collected in different parts of the world. With regard to the earth's antiquity he concluded that "no more definite statement can . . . be made at present than that the age of the Earth exceeds 1460 million years, is probably not less than 1600 million years, and is probably much less than 3000 million years" (*18*).

Time Scales (millions of years to base) 1961–1975 (19)

	Kulp *1961*	Harland *et al.* *1964*	Lambert *1971*	Armstrong & McDowell *1974*	Afanas'yev & Zykov *1975*
Cenozoic	63	70	65	65	66
Cretaceous	135	136	135	143	132
Jurassic	181	193	200	212	185
Triassic	230	225	240	247	235
Permian	280	280	280	289	280
Carboniferous	345	345	370	367	345
Devonian	405	395	415	416	400
Silurian	425	435	445	446	435
Ordovician	500	500	515	509	490
Cambrian	600	570	590	575	570

Holmes' estimates have been stretched by more recent dating of Precambrian rocks. Samples from southwestern Greenland have yielded radiometric ages of 3,760,000,000 years. But the earth must be even older, though how much so is still a subject of speculation. Lunar rocks are giving ages approaching 4,600,000,000, which is in the same range as ages of certain meteorites.

Meanwhile the calibration of the standard stratigraphic column continues as geologic investigations turn up new supplies of dateable materials and as techniques of radiometric dating are refined. The table shows determinations made by different investigators since 1960 when Holmes published his last figures.

Doubtless these figures will change in the light of new information, but their ranges suggest that the changes are unlikely to be spectacular. Indeed, all but three of the most recent determinations fall within the ranges given by Holmes fifteen years earlier.

CHAPTER SEVENTEEN

World Enough and Time

Physics, beware of metaphysics.
ISAAC NEWTON

The concept of geologic time, as expanded since the seventeenth century, has become a common denominator in the theory of all the historical sciences. The idea of an earth several billions of years old is the least astonishing to the astronomers, who measure their distances in light-years, and whose estimates of the time required for the unfolding of the universe are greater still. This same idea is also essential to organic evolution, the unifying concept in the life-related sciences of embryology, comparative anatomy, ethology (the science of behavior), biochemistry, parasitology, biogeography, paleontology, and archaeology.

That the earth must be ancient almost beyond comprehension is a concept which grew from the work of the early uniformitarians—Hutton and his followers. From the present composition, structure, and configuration of the earth's crust they inferred long series of past changes in these attributes that have led up to the present state. To their credit, they attempted to account for all past changes in terms of agencies operating today. Supernatural causes were ruled out, except for a "first cause" of creation. Under their influence the older speculations that the Alps may have been heaved up in the course of a single earthquake and that the great canyons of the world were carved by a single flood faded away.

Hindsight shows, however, that Lyell's double-barrelled uniformity principle had its defects. The assumption that past changes have always transpired at about the same average rate as those in progress today is admittedly gratuitous, as some of his critics protested. The companion proposition that there should be no appeal to causes of change other than those now in action is still useful as describing a method of investigation, but even so must be used with caution. For example, the earth has not in our memory been bombarded with giant meteorites of the kind that produced Meteor Crater of Arizona. Yet there are anomalous subcircular and strongly deformed structures in the older rocks which almost certainly record the impact and explosion of meteorites. For many years geologists sought to explain the Vredefort Dome of Africa and similar structures elsewhere in terms of volcanic action—a process demonstrably in operation now. Only after these efforts failed to account for the evidence was the meteoritic hypothesis seriously considered. The procedure was correct. Methodologic uniformity can be salvaged if restated as a geological corollary of the logical principle of simplicity (Occam's razor). Causes of geological change should not be multiplied without necessity, we can say. In which case the principle is common to all science and not peculiar to historical geology (*1*).

Two surprising facts have emerged from modern geochronology. Most of time past was spent before the beginning of the Cambrian. And the family of man appeared on the scene much longer ago than could have been predicted at the beginning of our century.

Investigations of prehistory are disclosing so much new information about ancient man that anything written on the subject is likely to be outdated before the printing. For example, explorations in northern Tanzania conducted in 1978 have turned up a trail of human footprints impressed on volcanic ash erupted from Mt. Sadiman upon the Laetoli Plain. A potassium-argon date of 3.6 million years was obtained for the ash overlying the prints, and one of 3.8 million for ash below. The prints belong to two individuals. On the bases of the length of the footprints and the stride, the smaller of the two stood about five feet tall, and the larger about eight inches taller. Judging from associated human teeth and bones, the prints were made by *Australopithecus afarensis*. Some sixty feet of ash and other sediment had covered the tracks before they were partially exposed by recent stream erosion (*2*).

These traces of early hominids seem almost unbelievably ancient, until we place them where they belong on the scale of geologic time. Imagine that the some 4,500 million years of earth history are compressed into the frame of one year's time. Each of the 365 days would then stand for 12,300,000 years, each hour for 513,000 years, each minute for 8,550 years, and each second for 142 years. On that basis, shell-producing marine invertebrates did not become abundant (with the beginning of the Cambrian) until around November 16. Vertebrate animals, the primitive fishes of the early Ordovician, appeared on the scene about November 21; and before the end of the month amphibians had established themselves as the first terrestrial vertebrates. Reptiles began to dominate the land about December 7, and mammals appeared around December 14, the first birds on the following day. Primates were present on Christmas Eve. Around five in the afternoon of New Year's Eve, the two hominids walked across the Laetoli Plain and left their prints behind. *Homo sapiens* appeared about an hour before midnight—or some 500,000 years ago.

On this same comparative scale, plants resembling modern blue-green algae were on the scene as early as May 2, compared with November 6 for multicellular animal life. The first vascular plants appeared around November 27, and forests of tree ferns, club mosses, and horsetails had covered parts of the continents when the amphibians began to occupy the lands around November 30. Flowering plants bloomed late, around December 21, and grasses appeared three days later.

In panoramic view the fossil record leaves no doubt that simple and one-celled organisms preceded multi-celled and complex organisms in time. Moreover, the major categories of *phyla*, into which zoologists classify animals, had appeared before the end of the Ordovician. The fundamental diversities manifest in the animal kingdom are thus of long standing, though in terms of our comparative calendar they were not accomplished until the fall of the year. Even so, the more than two million species of organisms now living are regarded by biologists as the culmination of evolution from a few parent stocks through the species-creating process of natural selection.

Natural selection is no longer regarded as a bloody struggle waged by individuals, only the fittest of which manage to survive. Evolution involves populations, and fitness involves different degrees of success among their members in producing numbers of viable offspring. Within the populations of each species, some individuals produce more offspring than others, and if these fecund ones differ genetically from those less successful at reproduction, their traits will tend to become dominant in the genetic pool of the population. Resulting accentuations of genetic differences over many generations may produce new varieties and species, although the record clearly shows that more often than not extinction is the ultimate reward for such progress (*3*).

Any emphasis given here to the importance of radiometric dating should not detract from the significance of ordering geologic events on a relative rather than numerical scale of time. In their daily operations, historical geologists still rely upon the principles of superposition, horizontality, original continuity and faunal sequence. A fourth self-evident proposition states that fractures and crosscutting bodies of rock must be younger than whatever they transect. With the aid of these principles rocks, structures, and fossils are ordered in chronologic sequence. Any dates in years that may be obtained for individual events are then fitted into the sequence. If all we knew about earth history were confined to a map of the globe showing numbers at points where numerical dates were derived, we should know very little. The relative time scales and the numbers for ages of minerals and rocks work together in a self-correcting system.

Now, as in Darwin's day, evolution is under attack. Some opponents argue that since evolutionary theory is contrary to any literal interpretation of Scripture it must be not only wrong but atheistic too. At the opposite pole are those who perceive evolution as a religious doctrine in its own right. Beginning with either interpretation, the last line in the argument turns out to be the same: evolution should not be taught in public schools; but if taught it should be presented in parallel with creationism, so that students may make free choices between alternatives.

The issue of *Science* for June 1, 1979, carries notice of a suit brought by a creationist group to have all exhibits on evolution removed from the American Museum of National History.

> The suit alleges that the theory of evolution is no more subject to scientific verification as an explanation for the origin of man and life on earth than the biblical story of creation, and that to believe in the one in-

volves just as much of an act of faith as to believe in the other. The religion of the evolutionists, the plaintiffs say, is "secular humanism." (*4*)

Earlier this year the Education Committee of the Minnesota House of Representatives rejected a bill that would have required schools to teach the evolution theory and the creation theory "with reasonably equal emphasis." Under the provisions of the bill, teachers would have been prohibited from lowering the grade of any student who preferred one theory to the other (*5*).

Creationist writings designed for use in American high schools, mostly as texts or supplementary reading in biology classes, necessarily get involved with historical geology. Favorite strategies for discrediting evolutionary biology concentrate on shortening geologic time to the point that natural selection couldn't have worked, and on attempting to demonstrate that man co-existed with various extinct forms of life that flourished during the Paleozoic and Mesozoic eras.

In order to shrink geologic time from the billions back into the thousands of years required by Biblical chronicles, the creationists resort to catastrophism. For example, the mile-thick sequence of sedimentary rocks exposed along the walls of the Grand Canyon has been interpreted as deposits formed during Noah's Flood, and the canyon itself as the work of back-rushing floodwaters. This erosion may have been unusually rapid, they suggest, because the sediments had not consolidated before the canyon was cut. But what of the orderly succession of fossils in the strata? That too is explained away, on the assumption that the simplest forms of life settled first from the Flood, while more advanced organisms escaped destruction longer, and so were entombed toward the top of the pile (*6*).

To demonstrate the antiquity of man on a time scale measured in thousands of years, creationists often refer to giant footprints impressed in Cretaceous limestone around Glen Rose, Texas. These tracks remind them of descriptions in the Bible of giant men who lived in the past. But as paleontologists, and indeed most of Glen Rose's citizens, have long known, the tracks were made by dinosaurs. In one creationist text is a picture of an elongate concretion split half in two, with trilobites inside. The concretion is interpreted as the imprint of a sandal worn by someone during Paleozoic time (*7*). What this individual was about, walking around on the ocean floor in sandals, is left to the reader's imagination.

One can only conclude that some creationists, recoiling from the fearsome prospect of time's abyss, have toppled backward into the abyss of ignorance.

> If daunted by the noxious stench
> Exhaled from Time's Abyss,
> Retreat into some lesser trench
> Where ignorance is bliss. (*8*)

Along with radical catastrophism, another idea current in the seventeenth century but long since abandoned is the conception of the world as a great ruin in process of final decay. The modern concept holds that while our planet is vastly ancient it is not senile. Endowed with powerful stores of internal energy, the earth continues to change in structure and configuration. The crust of the earth is demonstrably mobile: its crust deforms daily in response to the gravitational attraction of the moon, and is still bowing up in regions from which the loads of continental ice caps have recently melted away. The globe vibrates incessantly as seismic waves pass through its body; fortunately, most of these quakes are too small for detection except by delicate recording devices. Intermittently, as a result of volcanic activity, new mountains form and old ones grow by addition of lava, or less commonly are wrecked by explosions as at Krakatoa.

Throughout the first three decades of our century, influential textbooks advocated that the continents, regardless of their up-and-down movements leading to alternate episodes of emergence and submergence through time, have remained in their present positions relative to one another as far back as the record goes. This proposition was challenged in 1915 and afterward by the German meteorologist Alfred Wegener (*9*). The neat "fit" between the eastern coast of South America and the western coast of Africa was only the starting point in an array of evidence Wegener assembled in support of his continental drift theory. In his view, all present continents were joined as one at some time in the past and have since drifted apart. At first his theory was a subject of derision and its advocates were sometimes dismissed as "drifters." In the light of recent investigations, however, most geologists are receptive to the idea that the continents have shifted their positions laterally as well as up or down.

When the concept of the permanence of continents and ocean basins dominated the thinking of geologists, it seemed possible that somewhere in the depths of the oceans might be found an unbroken sequence of sediments and fossils extending back to early stages in the earth's history. That hope has dimmed because of results from coring sediments and rocks from the bottom of the ocean. No rocks older than the Jurassic have been recovered from cores taken from many parts of the world. And yet there is compelling evidence that the oceans existed long before the beginning of the Cambrian Period.

Studies of the sinuous ridge of volcanic sea-mountains that constitute the Mid-Atlantic Ridge suggest that the ridge marks a crack in the earth's crust, away from which the Americas have moved westward, and the African and Eurasian continents have moved eastward, since a breakup during the late Triassic. The Mid-Atlantic Ridge is but one segment in a network of global lineaments that divide the crust into plates. And the movement of these plates relative to one another is held accountable for the dispersal of the

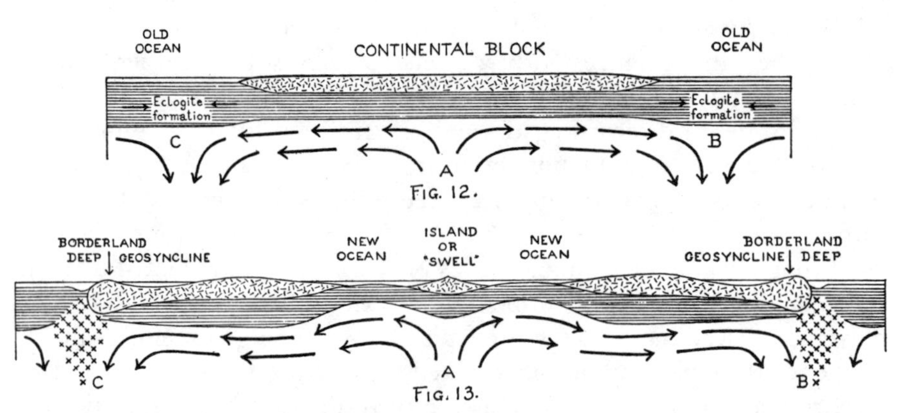

Fig. 12.—Sub-continental circulation in the Substratum with eclogite formation from the basaltic rocks of the Intermediate Layer above B and C where the sub-continental currents meet sub-oceanic currents and turn down. Upper "Granitic" Layer, dotted. Intermediate Layer (gabbro, amphibolite, basic granulite, etc.), line-shaded. Substratum, unshaded.

Fig. 13.—Distension of the continental block on each side of *A* with formation of new ocean floors from rising basaltic magma. The front parts of the advancing continental blocks are thickened into mountainous borderlands with oceanic deeps in the adjoining ocean floor due to the accumulation of eclogite at B and C.

Diagrams by Arthur Holmes (1928) illustrating his conception of how crustal plates might be fragmented and rafted apart by slow-moving convection currents in subcrustal materials. (Reproduced from the original drawings by Mrs. Holmes, courtesy Donald B. McIntyre) present continents away from the original continent. Slow-moving convection currents in the subcrustal region are commonly thought to be the agents that raft the plates about.

Whatever the mechanism, the idea of the dispersal of the continents solves some sticky problems raised by the older concept of permanently placed continents. For example, to explain the presence of the same kinds of land plants and land animals on opposite sides of the present oceans, geologists have sometimes postulated that land bridges formerly connected the continents, served as lanes for migrations of terrestrial organisms, and then sank and disappeared. These hypothetical bridges become unnecessary if the continents were joined at the times the species in question dispersed. This is only one type of evidence supporting the theory of sea-floor spreading that has grown out of Wegener's speculations. The details are not so important in the present context as the fact that the ruling theory in modern structural geology demands a mobility of the "solid earth" beyond anything dreamed at the beginning of this century.

Here again geologic theory is anchored to the concept of the lengthy duration of geologic time. The presumed separation of the Americas from continental masses to the east could have been accomplished by an average annual rate of separation amounting to no more than about four inches per year. But as Lyell maintained, a series of small directional changes adds up to spectacular results, given enough time. And now it appears that there's world enough and time to accommodate compatible, if not yet unified, theories for the chemical, physical, and biological evolution of the earth.

NOTES

CHAPTER TWO. *Malta's Cradle*

1. Springer. **2.** Young, vol. 2. **3.** Scherz, Gustav, 1969, Niels Stensen's geological work, pp. 11–47 *in* Scherz (1969). **4.** Steno, Nicolaus, 1660, Disputatio physica de thermis (A scientific disputation on the subject of hot springs); English translation of the Latin text by Alexander J. Pollock; pp. 49–63 *in* Scherz (1969). **5.** Steno, 1669. A facsimile of the French text was published in Copenhagen by Arnold Busck in 1965. To this is appended an English translation by Alexander J. Pollock (with introduction by Gustav Scherz) and a German translation by Adolf Pilz. **6.** Steno, 1667. **7.** Adams, 112–117. **8.** Plinius, Book 37, p. 627. **9.** Acts of the Apostles, Chapt. 28, verses 1–6. **10.** Zammit-Maempel. **11.** Hall and Hall, vol. 2, p. 121. Quoted with permission of the University of Wisconsin Press. **12.** Hall and Hall, vol. 5, p. 90. Quoted with permission of the University of Wisconsin Press. **13.** Garboe (1958). Quoted with permission of St. Martin's Press.

CHAPTER THREE. *Solids within Solids*

1. Sketches of Steno's life are found in Winter, pp. 169–187, and Scherz, 1969, pp. 11–47. **2.** For an English translation, with notes and commentary, *see* Winter, 1916. Both the Latin and English versions, also with notes and commentary, are given in Scherz, 1969, pp. 134–234. **3.** Winter, p. 210. **4.** Winter, p. 230. **5.** Winter, p. 230. **6.** Winter, p. 230. **7.** Winter, p. 272 and Pl. 9. **8.** Winter, pp. 262–270. **9.** Winter, p. 271. **10.** Scherz, 1960, p. 17. **11.** Oldenburg, 1671. **12.** Eyles, V. A., 1958, The influence of Nicolaus Steno on the development of geological science in Britain, pp. 167–189 *in* Scherz, ed., 1958. **13.** Scherz, Gustav, 1958, Nicolaus Steno's life and work, pp. 9–86 *in* Scherz, ed., 1958. **14.** Bacon, 1857, vol. 3, p. 351. **15.** Kuhn, p. 5. **16.** Kuhn, pp. 52–53. **17.** Kuhn, pp. 78–79. Quoted with permission of Thomas S. Kuhn and the University of Chicago Press.

CHAPTER FOUR. *Mr. Hook(e)*

1. Roy. Soc. London, Phil. Tr., vol. 2, 1667/8, p. 628. **2.** A contemporary account of Hooke's life and works by Richard Waller was published as an introduction to *The posthumous works of Robert Hooke, M. D.* (1705). R. T. Gunther has reproduced Waller's text, with inserts of additional biographical materials by John Ward and John Aubrey, in pp. 1–68 of his *Early Science in Oxford*, vol. 6, pt. 1, 1930. Robinson and Adams include a biographical sketch in their introduction to Hooke's diary of 1672–1680. 'Espinasse's *Robert Hooke* is not a biography in the usual sense, but more an analysis of Hooke's character and an appraisal of the scientific contributions of "this brilliant, generous, unlucky man." **3.** For an English translation of the Charter, *see* Sprat, 1667, pp. 134–143. **4.** This is the full title of the work cited in the References as Hooke, 1705. **5.** Hooke, p. 290. **6.** Hooke, p. 317. **7.** Hooke, p. 291. **8.** Hooke, p. 335. **9.** Rossiter, 1935, pp. 179–180. **10.** For a modern geological interpretation of the Atlantis story, *see* Vitaliano, 1973, pp. 218–251. **11.** Hooke, p. 324. **12.** Hooke, pp. 313–314. **13.** Hooke, p. 404. **14.** Hooke, p. 422. **15.** Hooke, pp. 338–343. **16.** Hooke, p. 410. **17.** Hooke, pp. 423–424. **18.** Hooke, pp. 427–428. **19.** Hooke, p. 435. **20.** Hooke, p. 326. **21.** Hooke, p. 348. **22.** Hooke, pp. 426–427. **23.** Hooke, p. 435. **24.** Waller, pp. xxvi–xxvii. **25.** Robinson and Adams, 1935, pp. 463–470. **26.** Robinson and Adams, 1935; entries for April 7, May 18, 19, 25, and 31, 1675. **27.** Robinson and Adams, 1935, pp. 205–207. Quoted with permission of Taylor and Francis Ltd. **28.** For references to Hooke in Pepys' diary, *see* Pepys, 1946, vol. 2, p. 1151. **29.** Pepys, 1920, vol. 4, p. 428. **30.** Pepys, 1946, vol. 1, p. 1036. **31.** 'Espinasse, 1956, pp. 141–155.

CHAPTER FIVE. *A Sacred Theory of the Earth*

1. *Telluris Theoria Sacra*, published in London in 1681, consisted of two books. These were translated into English and issued under the title *The Theory of the Earth: Containing an Account of the Original of the Earth, and of all the General Changes Which it hath already undergone, or is to undergo, Till the Consummation of all Things: The Two First Books Concerning The Deluge and Concerning Paradise* (London, Walter Kettilby, 1684). The third and fourth books, *Concerning the Burning of the World, and Concerning the New Heavens and New Earth*, were issued by the same publisher in 1690. A 1691 edition of the first two books combined with the 1690 edition of the third and fourth was reprinted in 1965 by the Southern Illinois Press with an introduction by Basil Willey. Most page references to *Sacred Theory* cited below refer to the Southern Illinois reprint. **2.** An extended account of *Sacred Theory* and the Burnet controversy is given in Nicolson, 1963, pp. 184–270. **3.** Burnet, 1965, p. 16. **4.** Burnet, 1965, p. 34. **5.** Burnet, 1965, p. 38. **6.** Burnet, 1965, p. 45. **7.** Burnet, 1965, pp. 51–53. **8.** Burnet, 1965, p. 57. **9.** Burnet, 1684, pp. 67–68. **10.** Burnet, 1965, pp. 82–88. **11.** Burnet, 1965, p. 91. **12.** Burnet, 1684, pp. 109–110. **13.** Burnet, 1965, p. 281. **14.** Burnet, 1965, p. 91. **15.** Warren, 1960, p. 143. A critical appreciation of Warren's writing is given by Nicolson, 1963, pp. 263–269. **16.** Sprat, p. 417. **17.** King, pp. 121–122. **18.** Burnet, 1965, p. 218.

CHAPTER SIX. *Telliamed's Story*

1. A translation by A. V. Carozzi of J. B. le Mascrier's biographic sketch of de Maillet is given on pp. 407–413 of de Maillet, 1968. Almost all the materials in this chapter are based on the Carozzi edition of *Telliamed.* **2.** Carozzi gives a detailed account of Cartesian cosmogony in de Maillet, 1968, pp. 32–34. **3.** Carozzi, *in* de Maillet, 1968, p. 34. **4.** de Maillet, 1968, pp. 183–189. **5.** de Maillet, 1968, pp. 191–200. **6.** de Maillet, 1968, pp. 126–130. **7.** de Maillet, 1968, pp. 130–132. **8.** de Maillet, 1968, pp. 122–123. **9.** de Maillet, 1968, pp. 92–93. **10.** de Maillet, 1968, p. 181. **11.** Carozzi, *in* de Maillet, 1968, pp. 5–10. **12.** Carozzi, *in* de Maillet, 1968, pp. 15–26. **13.** Carozzi, *in* de Maillet, 1968, pp. 27–30. **14.** Carozzi, *in* de Maillet, 1968, pp. 3, 50. **15.** Adams, 1938, pp. 209–276. **16.** de Maillet, 1968, pp. 160–162. **17.** de Maillet, 1968, pp. 96–97. **18.** Carozzi, *in* de Maillet, 1968, p. 4.

CHAPTER SEVEN. *The Epochs of Nature*

1. A chronology for publication of the successive volumes of *Natural History* is given on pp. 11–13 of Fellows and Milliken, 1972. **2.** Fellows and Milliken, 1972, p. 41. **3.** Fellows and Milliken, 1972, pp. 47, 51–54. **4.** Fellows and Milliken, 1972, p. 48. **5.** The following account of Buffon's cosmology, geology, biology, and method of investigation derives from Jacques Roger's excellent critical commentary, pp. xliii–xciv *in* Buffon, 1962. **6.** Roger (*in* Buffon, 1962) pp. lx–lxv. **7.** Roger (*in* Buffon, 1962) p. lxv. **8.** Fellows and Milliken, 1972, p. 16. **9.** Roger (*in* Buffon, 1962) p. lxxxviii. **10.** Fellows and Milliken, 1972, p. 56. **11.** Roger (*in* Buffon, 1962) pp. lxxxvii–lxxxviii. **12.** Fellows and Milliken, 1972, p. 23. **13.** Buffon, 1962, pp. 19–24. **14.** Fellows and Milliken, pp. 82–83. **15.** Quoted by Roger (*in* Buffon, 1972) p. xxvii.

CHAPTER EIGHT. *View from the Brink*

1. Playfair, 1805, pp. 40–41. **2.** Playfair, 1805, pp. 43–44. **3.** Playfair, 1805, p. 44. **4.** Playfair, 1805, p. 45. **5.** Playfair, 1805, p. 51. **6.** Playfair, 1805, p. 51. **7.** Page references to the essay refer to its reproduction as Chapter I in Hutton, 1795. **8.** Hutton, 1795, vol. 1, p. 17. **9.** Hutton, 1795, vol. 1, pp. 42–49. **10.** Hutton, 1795, vol. 1, pp. 53–59. **11.** Hutton, 1795, vol. 1, p. 118. **12.** Hutton, 1795, vol. 1, p. 146. **13.** Hutton, 1795, vol. 1, p. 200. **14.** Hutton, 1795, vol. 1, p. 271. **15.** Hutton, 1795, vol. 1, pp. 271–272. **16.** Hutton, 1795, vol. 1, p. 127. **17.** Bailey, 1967, p. 65. **18.** Bailey, 1967, pp. 64–67; 69–73. **19.** Hutton, 1795, vol. 1, pp. 222–223. **20.** Playfair, 1805, p. 93. **21.** Playfair, 1805, p. 98. **22.** Playfair, 1805, p. 91. **23.** Playfair, 1805, pp. 68–69. **24.** Hutton, 1795, vol. 2, pp. 222–223. **25.** Hutton, 1795, vol. 2, p. 236. **26.** Hutton, 1795, vol. 1, p. 374. **27.** Hutton, 1795, vol. 1, p. 273. **28.** Hutton, 1795, vol. 1, p. 297. **29.** Hutton, 1795, vol. 2, p. 562. *See also* Tomkeieff, 1949, pp. 396–397; and McIntyre, 1963. **30.** Hutton, 1795, vol. 2, p. 239. **31.** Hutton, 1795, vol. 1, pp. 454–465. **32.** Playfair, 1805, pp. 71–72. **33.** For a perceptive analysis of Hutton's philosophy *see* O'Rourke, 1978.

CHAPTER NINE. *Father William*

1. Schenck, 1961, pp. 3–4. **2.** Rudwick, 1976, pp. 127–128. **3.** For biographical accounts of Smith *see* Cox, 1942 and 1948; J. M. Eyles, 1968-a; and Phillips, 1844. A bibliography of Smith's published works is given by J. M. Eyles, 1969-b. **4.** Cox, 1942, p. 10. **5.** Phillips, 1844, p. 29. **6.** Sheppard, 1917, pp. 109–112. **7.** Smith, 1806; Sheppard, 1917, pp. 112–117. **8.** Sheppard, 1917, pp. 109–112. **9.** Smith, 1815-a. **10.** Smith, 1815-b, pp. 2–3. **11.** Smith, 1816–1819. **12.** Smith, 1817. **13.** Eyles, J. M., 1969-a, p. 157. **14.** Porter, 1977, pp. 146–148. **15.** Woodward, 1907, pp. 91–92. **16.** Adams, 1938, p. 275. **17.** Cox, 1942, p. 5. **18.** Smith, 1817, p. vi. **19.** Quoted by Bassett, 1969, p. 40. **20.** Smith, 1817, p. vi. **21.** Smith, 1817, p. x. **22.** North, 1927. **23.** Smith, 1817, p. x. **24.** Do. **25.** Quoted by Cox, 1942, p. 70.

CHAPTER TEN. *The Four-dimensional Jigsaw Puzzle*

1. An account of Lehmann's life and works is given in Freyberg, 1955. **2.** Adams, 1938, pp. 372–375. **3.** For a sketch of Werner's life and contributions to historical geology, *see* the introduction by Ospovat in his translation of Werner's *Short Classification and Description of the Various Rocks*. **4.** Werner,

1774. *See* the Carozzi translation of 1962 for the significance of this book in mineralogy. **5.** Werner, 1786. *See* Ospovat translation of 1971. **6.** Ospovat, 1969, p. 3 ff. **7.** Quoted by Adams, 1938, pp. 220–221. **8.** Werner, 1786, p. 44 (Ospovat translation). **9.** Do., p. 20. **10.** White, 1977, p. 267. **11.** An extended account of the development of the standard stratigraphic column and scale of geologic time is given by Berry, 1968. **12.** Murchison, 1835. **13.** Sedgwick and Murchison, 1836. **14.** Lapworth, 1879. **15.** Sedgwick and Murchison, 1839. **16.** Murchison, 1841. **17.** Moore and Slusher, 1976, p. xix. **18.** Cited by Bowler, 1976, p. 37.

CHAPTER ELEVEN. *A Question of Tempo*

1. Quoted by Woodward, 1907, p. 85. **2.** Woodward, 1907, p. 44. **3.** Cuvier, 1817. **4.** Cuvier, 1817, pp. 15–17. **5.** Cuvier, 1817, pp. 17–18. **6.** Culvier, 1817, pp. 22–23. **7.** Cuvier, 1817, pp. 171–172. **8.** Cuvier, 1817, p. 163. **9.** Cuvier, 1817, pp. 180–181. **10.** Jameson, *in* Cuvier, 1817, p. 221. **11.** Most of the biographical information on Lyell in this essay is from the excellent biography by Wilson, 1972. **12.** Lyell, 1830, pp. 61–65. **13.** Lyell, 1872, p. 97. **14.** Do. **15.** Lyell, 1872, p. 99. **16.** Lyell, 1830, pp. 78–79. **17.** Lyell, 1872, p. 90. **18.** Lyell, 1872, p. 91. **19.** Lyell, 1872, p. 93. **20.** Lyell, 1830, p. 190. **21.** Lyell, 1851, p. xxxii. **22.** For an analysis of Lyell's methodology, *see* Rudwick, 1970. **23.** Sedgwick, 1831, p. 301. **24.** Sedgwick, 1831, p. 306. **25.** Sedgwick, 1831, p. 315. **26.** Sedgwick, 1831, p. 307. **27.** Sedgwick, 1831, p. 314. **28.** Sedgwick, 1831, pp. 315–316. **29.** Sedgwick, 1831, p. 299. **30.** Whewell, 1831, p. 202.

CHAPTER TWELVE. *A Plenitude of Events*

1. A summary account of early geological investigations in the Auvergne is given by Scrope, 1858, pp. 30–39. **2.** Adams, 1938, p. 245. **3.** Scrope, 1858, p. 206. **4.** Scrope, 1858, p. viii. **5.** Scrope, 1858, p. 205. **6.** Scrope, 1858, pp. 208–209. **7.** Marcou, 1896, vol. 1, p. 7. **8.** For a summary account of the forerunners of the glacial hypothesis, *see* Carozzi, 1967, pp. xi–xvii. **9.** Marcou, 1896, vol. 1, pp. 74–75. **10.** Marcou, 1896, vol. 1, p. 75. **11.** Marcou, 1896, vol. 1, p. 73. **12.** Marcou, 1896, vol. 1, p. 85. **13.** For English translation and commentary *see* Carozzi, 1967, pp. xliii–lxi. **14.** Carozzi, 1967, p. liv. **15.** Carozzi, 1967, pp. liv–lvii. **16.** Carozzi, 1967, p. lviii. **17.** Marcou, 1896, vol. 1, pp. 109–110. **18.** Carozzi, 1967, p. 136.

CHAPTER THIRTEEN. *The Great Tree of Life*

1. de Beer, p. 12. **2.** de Beer, p. 28. **3.** de Beer, p. 33. **4.** Barlow, pp. 4–6. **5.** de Beer, p. 36. **6.** Barlow, p. 30. **7.** de Beer, p. 44. **8.** Barlow, pp. 136–140. **9.** de Beer, p. 58. Colp, pp. 139–144, believes that Darwin's illness was more probably related to psychological stress than to infectious causes. **10.** de Beer, p. 72. **11.** de Beer, p. 72. **12.** Peckham, p. 9. **13.** Darwin, 1878, vol. 1, p. 34. **14.** Darwin, 1878, vol. 1, p. 79. **15.** Darwin, 1878, vol. 1, pp. 79–80. **16.** Darwin, 1878, vol. 1, pp. 77–78. **17.** Darwin, 1878, vol. 1, p. 96. **18.** Darwin, 1878, vol. 1, pp. 3–4. **19.** Darwin, 1878, vol. 2, pp. 51–52. **20.** Darwin, 1878, vol. 1, p. 118. **21.** Darwin, 1878, vol. 2, p. 76. **22.** Darwin, 1878, vol. 1, p. 99. **23.** Darwin, 1878, vol. 1, pp. 162–163.

CHAPTER FOURTEEN. *Kelvin*

1. Kelvin, 1852, p. 306. **2.** For an excellent and comprehensive account of Kelvin's investigations concerning the age of the earth, *see* Burchfield, 1975. **3.** In this chapter most of the biographical information on Kelvin is from Thompson's two-volume work of 1910. **4.** Thompson, 1910, pp. 32–58. **5.** Thompson, 1910, p. 162. **6.** For reproductions of certain letters of recommendation *see* Thompson, 1910, p. 167 ff. **7.** Thompson, 1910, p. 189. **8.** A list of Kelvin's awards, his bibiography, and a listing of his patents are appended to Thompson, 1910, vol. 2. **9.** Thompson, 1910, p. 914 (reproduction and description of arms). **10.** Kelvin, 1869, p. 476. **11.** Kelvin, 1871, p. 17. **12.** Kelvin, 1871, p. 18. **13.** *See* Thompson, 1910, pp. 537–538. **14.** Kelvin, 1871, pp. 18–19. **15.** Kelvin, 1899, p. 89. **16.** Kelvin, 1864, p. 161. **17.** Kelvin, 1871, p. 25. **18.** Kelvin, 1899, p. 75. **19.** Kelvin, 1871, pp. 5–16. **20.** Kelvin, 1899, pp. 71–73. **21.** Kelvin, 1899, p. 89. **22.** For a summary of the address and an account of the reception, *see* Thompson, 1910, pp. 598–610. **23.** Thompson, 1910, p. 610. **24.** Thompson, 1910, p. 1103. **25.** Kelvin, 1899, p. 89. **26.** For an extended account of Kelvin's views on religion and other matters, *see* Thompson, 1910, 1086–1146.

CHAPTER FIFTEEN. *A Numbers Game*

1. For a detailed account of Kelvin's influence and the reactions to his ideas, *see* Burchfield, 1975, chapters 3–5. **2.** Kelvin, 1864, p. 159. **3.** Quoted by Kelvin, 1871, pp. 1–2. **4.** Kelvin, 1871, p. 16. **5.** Kelvin,

1871, p. 25. **6.** Huxley, 1869, p. xlvi. **7.** Huxley, 1869, p. xlvi. **8.** Huxley, 1869, p. l. **9.** Huxley, 1869, p. liii. **10.** Kelvin, 1869, p. 475. **11.** Tait, 1869 (published anonymously). The quotation is from Friedrich Engels' *Dialectics of Nature* (*see* Ebison, 1977, p. 54). **12.** For an account of Phillips' work and influence, *see* Burchfield, 1975, pp. 59–60 and other references. **13.** Morgan, 1878, p. 205. **14.** A synopsis of Wallace's method is given by Upham, 1893, pp. 211–212. **15.** A table comparing 19 such estimates (1860–1909) is given in Eicher, 1978, p. 14. **16.** McGee, 1893, p. 310. **17.** McGee, 1893, p. 310. **18.** Perry, John, 1895-a. **19.** Tait, quoted in Perry, 1895-a, p. 226. **20.** Kelvin, quoted in Perry, 1895-a, p. 227. **21.** Perry, 1895-b, p. 585. **22.** Fisher, 1893. **23.** Fisher, 1882. **24.** Fisher, 1882. **25.** Chamberlin, 1899, p. 890. **26.** Chamberlin, 1899, p. 12. **27.** Sollas, 1900, p. 291. **28.** H. B. Woodward, 1911, p. 188.

CHAPTER SIXTEEN. *Radiometric Dating*

1. For an account of Rutherford's life and work, *see* Eve, 1939. **2.** Quoted by Holmes, 1931, p. 130. **3.** Rutherford, 1905, p. 780 (Collected Papers). **4.** Rutherford, 1904, p. 657. **5.** Strutt, 1905, p. 99. **6.** Boltwood, 1907. **7.** Thompson, 1910, vol. 2, p. 1097. **8.** Eve, p. 107. Quoted with permission of Cambridge University Press. **9.** Reade, 1906, p. 79. **10.** For a detailed account of the geological reaction to expanded ranges of geologic time, *see* Burchfield, 1975, pp. 179–190. **11.** Holmes, 1911-a, p. 256. **12.** For a brief account of Holmes' life and career, *see* Dunham, 1966. **13.** Holmes, 1911-b, p. 9. **14.** Adapted from Wager, 1964, Fig. 5. **15.** See Cloud, 1978, p. 62. **16.** Schuchert, 1931, pp. 14–15. **17.** Knopf, 1931, p. 71. **18.** Holmes, 1931, p. 454. **19.** Adapted from Armstrong, 1978, Table 6, p. 90.

CHAPTER SEVENTEEN. *World Enough and Time*

1. Goodman, 1967. **2.** Leakey, 1979. **3.** For an account of modern evolutionary theory designed for the general reader, *see* Simpson, 1964. **4.** Carter, D. J., *Science*, vol. 204, p. 925, 1 June 1979. Copyright 1979 by the American Association for the Advancement of Science. **5.** Anonymous, 1979. **6.** Moore and Slusher, 1976, pp. 420–426. **7.** Moore and Slusher, 1976, p. 427. **8.** Graffito, wall of men's room, railway station at Capri. **9.** For an account of Wegener's theory and the theories developed from it, *see* Takeuchi *et al.*, 1970.

REFERENCES

Adams, Frank Dawson, 1938, The birth and development of the geological sciences: Baltimore, Williams & Wilkins

Agassiz, Louis, 1840, Études sur les glaciers (English translation and introduction by Albert V. Carozzi): New York and London, Hafner (1967)

Anonymous, 1979, Creation bill killed: Christian Century, May 9, p. 520

Armstrong, Richard Lee, 1978, Pre-Cenozoic Phanerozoic time scale—computer file of critical dates, and consequences of new and in-progress decay constant revisions, pp. 73–91 *in* Cohee, George V., *et al.*, eds., Contributions to the geologic time scale: Am. Assoc. Petroleum Geologists, Studies in Geology No. 6

Bacon, Francis, 1857, Collected works (Basil Montagu, ed.): Philadelphia, Parry and McMillan, 3 vols.

Bailey, Edward B., 1967, James Hutton—the founder of modern geology: Amsterdam, London, New York; Elsevier

Bakewell, Robert, 1813, An introduction to geology . . . : London, J. Harding

Barlow, Nora, ed., 1967, Darwin and Henslow, the growth of an idea: Berkeley and Los Angeles, Univ. California Press

Barrell, Joseph, 1917, Rhythms and the measurement of geologic time: Geol. Soc. Am., Bull., vol. 28, pp. 745–904

Bassett, Douglas A., 1969, William Smith, the father of English geology and of stratigraphy: Geology (Teachers of Geology Assoc., Bull.), vol. 1, pp. 38–51

de Beer, *Sir* Gavin, ed., 1974, Charles Darwin/Thomas Henry Huxley autobiographies: London, Oxford Univ. Press

Berry, William B. N., 1968, Growth of a prehistoric time scale, based on organic evolution: San Francisco and London, W. H. Freeman & Co.

Boltwood, Bertram B., 1907, On the ultimate disintegration products of the radio-active elements, Pt. 2, The disintegration products of uranium: Am. Jour. Sci., 4th ser., vol. 23, pp. 77–88

Bowler, Peter J., 1976, Fossils and progress: paleontology and the idea of progressive evolution in the nineteenth century: New York, Science History Publications

Buffon, George-Louis Leclerc, *Comte de*, 1962, Des époques de la nature; Jacques Roger, ed.: Paris, Muséum National d'Histoire Naturelle; Mém., n.s., sér C, vol. 10

Burchfield, Joe D., 1975, Lord Kelvin and the age of the earth: New York, Science History Publications

Burnet, Thomas, 1684–1690, The theory of the earth: containing an account of the original of the earth, and of all the general changes which it hath already undergone, or is to undergo till the consummation of all things: London, R. Norton for Walter Kittilby, 2 vols.

——, 1690–1691, The theory of the earth . . . : London, R. Norton for Walter Kittilby, 2 vols. (Reprinted: Southern Illinois Univ. Press, with introduction by Basil Willey, 1965)

——, 1699, Telluris theoria sacra . . . : Amsterdam, J. Wolters

Carozzi, Albert V., ed. and trans., 1967, Studies on glaciers, preceded by the Discourse of Neuchâtel, by Louis Agassiz: New York and London, Hafner

Carter, Luther J., 1979, Creationists sue to ban museum evolution exhibits: Science, vol. 204, p. 925

Chamberlin, Thomas Chrowder, 1899, Lord Kelvin's address on the age of the earth as an abode fitted for life: Science, n.s., vol. 9, June 30, pp. 889–901; July 7, pp. 11–18

Cloud, Preston, 1978, Cosmos, earth, and man: New Haven and London, Yale Univ. Press

Colp, Jr., Ralph, 1977, To be an invalid: the illness of Charles Darwin: Chicago and London, Univ. Chicago Press

Cox, L. R., 1942, New light on William Smith and his work: Yorkshire Geol. Soc., vol. 25, pp. 1–99

——, 1948, William Smith and the birth of stratigraphy: 18th Internat. Geol. Congress, 8 pp.

Cuvier, Georges, 1817, Essay on the theory of the earth, with mineralogical notes and an account of Cuvier's geological discoveries by Professor Jameson (Translation of Discours sur les révolutions), 3d. ed.: Edinburgh, W. Blackwood

Darwin, Charles, 1878, The origin of species by means of natural selection; or the preservation of favored races in the struggle for life: Akron, Ohio, Werner, 2 vols.

——, 1959, The origin of species: a variorum edition edited by Morse Peckham: Philadelphia, Univ. Pennsylvania Press

Davies, G. L., 1969, The earth in decay: a history of British geomorphology 1578 to 1878: New York, Science History Publications

de Maillet, Benoît, 1968, Telliamed, or conversations between an Indian philosopher and a French missionary on the diminution of the sea (Albert V. Carozzi, trans. and ed.): Urbana, Univ. of Illinois Press

Dunham, Kingsley, 1966, Arthur Holmes (1890–1965): Biographical Memoirs, Fellows of the Royal Society, vol. 12, pp. 291–310

Ebison, Maurice, 1977, Scientific quotations: the harvest of a quiet eye: New York, Crane, Russak & Co.

Eicher, Don L., 1978, Geologic time: Englewood Cliffs, N.J., Prentice-Hall, 2d ed.

'Espinasse, Margaret, 1956, Robert Hooke: Berkeley and Los Angeles, Univ. California Press

Eve, A. S., 1939, Rutherford: being the life and letters of the Rt. Hon. Lord Rutherford, O.M.: New York, Macmillan: Cambridge, Eng., Univ. Press

Eyles, Joan M., 1969-a, William Smith: some aspects of his life and work, pp. 142–158 *in* Toward a history of geology (Cecil J. Schneer, ed.): Cambridge, Mass., M.I.T. Press

——, 1969-b, William Smith (1769–1839): a bibliography of his published writings, maps, and geological sections, printed and lithographed: Society for the Bibliography of Natural History, Jour., vol. 5, pt. 2, pp. 87–109

Eyles, V. A., 1958, The influence of Nicolaus Steno on the development of geological science in Britain: Acta Historica Scientiarum Naturalium et Medicinalium, vol. 15, pp. 167–188

Faul, Henry, 1978, A history of geologic time: Am. Scientist, vol. 66, pp. 159–165

Fellows, Otis E., and Stephen F. Milliken, 1972, Buffon: New York, Twayne

Fisher, Osmond, 1882, On the depression of ice-loaded lands: Geol. Mag., ser. 2, vol. 9, p. 526

——, 1893, Rigidity not to be relied upon in estimating the earth's age: Am. Jour. Sci., 3rd. ser., vol. 145, pp. 209–220

Freyberg, Bruno *von*, 1955, Johann Gottlob Lehmann (1719–1767), ein Arzt, Chemiker, Metallurg, Bergmann, Mineraloge, und grundlegender Geologe: Erlangen, Erlangen Forsch., ser. B, vol. 1, pp. 1–159

Garboe, Axel, ed. and trans., 1958, The earliest geological treatise (1667) by Nicolaus Steno (Niels Stensen); translated from Canis carchariae dissectum caput, with introduction and notes: London, Macmillan; New York, St. Martin's Press

Goodman, Nelson, 1967, Uniformity and simplicity, pp. 93–99 *in* Albritton, Claude C., ed., Uniformity and simplicity; a symposium on the principle of the uniformity of nature: Geol. Soc. Am., Special Paper 89

Gunther, R. T., 1930–1938, The life and works of Robert Hooke; vols. 6, 7, 8, 10, and 13 of Early science in Oxford: Oxford, printed for the Author

Hall, A. Rupert, and Marie Boas Hall, eds. and translators, 1965–1973, The correspondence of Henry Oldenburg: Madison and Milwaukee, Univ. Wisconsin Press, 8 vols.

Holmes, Arthur, 1911-a, The association of lead with uranium in rock minerals, and its application to the measurement of geologic time: Royal Soc., Proc., ser. A, pp. 248–256

——, 1911-b, The duration of geological time: Nature, vol. 87, July 6, pp. 9–10

——, 1913, The age of the earth: New York, Harper

——, 1927, The age of the earth, an introduction to geological ideas: New York and London, Harper

——, 1931, Radioactivity and geological time, pp. 124–466 *in* Physics of the earth, pt. 4, The age of the earth: Washington, D.C., Nat'l. Research Council, Bull. 80

——, 1937, The age of the earth (new ed.): London, Nelson

——, 1947, The construction of a geological time scale: Geol. Soc. Glasgow, vol. 21, pp. 117–152

——, 1960, A revised geological time-scale: Edinburgh Geol. Soc., Trans., vol. 17, pt. 3, pp. 183–216

Hooke, Robert, 1705, Lectures and discourses of earthquakes and subterraneous eruptions . . . ; pp. 279–450 *in* The posthumous works of Dr. Robert Hooke (Richard Waller, ed.): London, Smith and Walford. Reprint edition, 1978, by Arno Press, Inc., New York

Hutton, James, 1788, Theory of the earth; or an investigation of the laws observable in the composition, dissolution, and restoration of land upon the globe: Royal Soc. Edinburgh, Trans., vol. 1, pt. 2, pp. 209–304

——, 1795, Theory of the earth with proofs and illustrations: London; printed for Cadell and Davies: Edinburgh, printed for William Creech, 2 vols.

Huxley, Thomas Henry, 1869, Anniversary address of the President: Geol. Soc. London, Quart. Jour., pp. xxxviii–liii

Jameson, Robert, 1976, The Wernerian theory of the Neptunian origin of rocks: New York, Hafner (facsimile reprint of Elements of geognosy, 1808)

Joly, John, 1899, An estimate of the geological age of the earth: Smithsonian Institution, An. Rept., year ending June 30, 1899, pp. 247–288

Kelvin, William Thomson, First Baron, 1852, On a universal tendency in nature to the dissipation of mechanical energy: Phil. Mag., ser. 4, vol. 4, pp. 304–306

——, 1864, On the secular cooling of the earth: Edinburgh Royal Soc., Trans., vol. 23, pp. 157–170 (paper read April 28, 1862)

——, 1869, Geological dynamics: Geol. Mag., vol. 6, pp. 472–476

——, 1871, On geological time: Geol. Soc. Glasgow, Trans., vol. 3, pt. 1, pp. 1–28 (paper read Feb. 27, 1868)

——, 1899, The age of the earth as an abode fitted for life: Phil. Mag., ser. 5, vol. 47, pp. 69–90 (1897 address before Victoria Institute, with numerous additions written at various times from June 1897 to May 1898)

King, William, 1776, The original works of William King: London, printed for the editor and sold by N. Conant, 2 vols.

Knopf, Adolph, 1931, Age of the ocean, pp. 65–72 in Physics of the earth, pt. 4, The age of the earth: Washington, D.C., Nat'l Research Council Bull. 80.

Kuhn, Thomas S., 1970, The structure of scientific revolutions: Chicago, Univ. of Chicago Press, Internat. Encyclopedia of Unified Science, vol. 2, no. 2, 2d. ed.

Lapworth, Charles, 1879, On the tripartite classification of the Lower Paleozoic rocks: Geol. Mag., n.s., vol. 6, pp. 1–15

Leakey, Mary D., 1979, Footprints in the sands of time: Nat'l. Geographic, vol. 155, pp. 446–457

Lyell, Charles, 1830–1833, Principles of geology: London, John Murray, 3 vols.

——, 1851, Anniversary address of the President: Geol. Soc. London, Quart. Jour., pp. xxv–lxxvi

——, 1872, Principles of geology: London, John Murray, vol. 1, 12th ed.

McGee, W. J., 1893, Note on the "Age of the earth": Science, vol. 21, no. 540, pp. 309–310

McIntyre, Donald B., 1963, James Hutton and the philosophy of geology, pp. 1–11 *in* Albritton, Claude C., ed., The fabric of geology: San Francisco, Freeman, Cooper & Co.

Marcou, Jules, 1896, Life, letters, and works of Louis Agassiz: New York and London, Macmillan, 2 vols.

Moore, John N., and H. S. Slusher, eds., 1976, Biology: a search for order in complexity: Grand Rapids, Mich., Zondervan Corp.

Morgan, C. Lloyd, 1878, Geological time: Geol. Mag., ser. 2, vol. 5, pp. 154–162, 199–207

Murchison, R. I., 1835, On the Silurian System of rocks: London and Edinburgh Phil. Mag. and Jour. Sci., 3rd. ser., vol. 7, pp. 46–52

——, 1841, First sketch of some of the principal results of a second geological survey of Russia, in a letter to M. Fischer: Phil. Mag., vol. 19, pp. 419–422

Nicolson, Marjorie Hope, 1963, Mountain gloom and mountain glory: the development of the aesthetics of the infinite: New York, W. W. Norton & Co.

North, F. J., 1927, "Deductions from established facts in geology," by William Smith; notes on a newly discovered broadsheet: Geol. Mag., vol. 64, pp. 532–540

Oldenburg, Henry, tr., 1671, The prodromus of a dissertation concerning solids naturally contained within solids, by Nicolaus Steno: London, Moses Pitt (advertisement and summary in Royal Soc., Trans., vol. 6, no. 72, pp. 2186–2190)

O'Rourke, J. E., 1978, A comparison of James Hutton's *Principles of knowledge* and *Theory of the earth:* Isis, vol. 69, pp. 5–20

Ospovat, Alexander, 1969, Reflections on A. G. Werner's "Kurze Klassifikation," pp. 242–256 *in* Schneer, Cecil J., ed., Toward a history of geology: Cambridge, Mass., and London, M.I.T. Press

Pepys, Samuel, 1920, Diary and correspondence of Samuel Pepys, F.R.S.: New York, Bigelow, Brown and Co., 4 vols.

——, 1946, The diary of Samuel Pepys, M.A., F.R.S.: New York, Random House, 2 vols.

Perry, John, 1895-a, The age of the earth: Nature, vol. 51, pp. 224–227

——, 1895-b, The age of the earth: Nature, vol. 51, pp. 582–585

Phillips, John, 1844, Memoirs of William Smith, LL.D.: London, John Murray

Playfair, John, 1802, Illustrations of the Huttonian theory of the earth: Edinburgh, Cadell and Davies

——, 1805, Biographical account of the late James Hutton, F.R.S. Edinburgh: Royal Soc. Edinburgh, Trans., vol. 5, pt. 3, pp. 39–99

Plinius Secundus, Gaius, 1601, The historie of the world; translated into English by Philemon Holland: London, Adam Islep

Porter, Roy, 1977, The making of geology: earth science in Britain, 1660–1815: Cambridge, Cambridge Univ. Press

Ray, John, 1693, Three physico-theological discourses: London, Sam Smith

Reade, T. Mellard, 1906, Radium and the radial shrinkage of the earth: Geol. Mag., ser. 5, vol. 3, pp. 79–80

Robinson, Henry W., and Walter Adams, eds., 1935, The diary of Robert Hooke, M.A., M.D., F.R.S., 1672–1680: London, Taylor and Francis

Roger, Jacques, *see* Buffon, 1962

Rossiter, A. P., 1935, The first English geologist—Robert Hooke (1635–1703): Durham Univ. Jour., vol. 29, pp. 172–181

Rudwick, Martin J. S., 1970, The strategy of Lyell's *Principles of Geology:* Isis, vol. 61, pp. 4–33

——, 1976, The meaning of fossils: episodes in the history of palaeontology, 2d. ed.: New York, Neale Watson Academic Pubs.

Rutherford, Ernest (Lord Rutherford of Nelson), 1904, The radiation and emanation of radium, pt. II: Technics, August issue, pp. 171–175 (collected papers, vol. 1, pp. 650–657)

——, 1904, Radio-activity: Cambridge, Cambridge Univ. Press

——, 1905, Radium—the cause of the earth's heat: Harper's Mag., Feb. issue, pp. 390–396 (collected papers, vol. 1, pp. 776–785)

——, 1962–1965, The collected papers of Lord Rutherford of Nelson: London, Allen and Unwin, 3 vols.

Schenck, Hubert G., 1961, Guiding principles in stratigraphy: Geol. Soc. India, Jour., vol. 2, pp. 1–10

Scherz, Gustav, ed., 1958, Nicolaus Steno and his indice: Copenhagen, Acta Historica Scientiarum Naturalium et Medicinalium, vol. 15, 314 pp.

——, 1960, Niels Stensen in Copenhagen: Geotimes, vol. 5, no. 1, pp. 10–17, 56–57

—— ed., 1969, Steno, geological papers (Alexander J. Pollock, tr.): Odense, Odense Univ. Press

——, 1969, Niels Stensen's geological work, pp. 11–47 in title cited above

——, ed., 1971, Dissertations on Steno as geologist: Odense, Odense Univ. Press

Schuchert, Charles, 1931, Geochronology, or the age of the earth on the basis of sediments and life, pp. 10–64 *in* Physics of the earth, pt. IV, The age of the earth: Washington, D.C., Nat'l. Research Council, Bull. 80

Scrope, G. Poulett, 1858, The geology and extinct volcanos of central France, 2d. ed.: London, John Murray

Sedgwick, Adam, 1831, Address to the Geological Society, delivered on the evening of the 18th of February: Geol. Soc. London, Pr., vol. 1 (1826–1833), pp. 281–316

Sedgwick, Adam, and R. I. Murchison, 1836, On the Silurian and Cambrian systems, exhibiting the order in which the older sedimentary strata succeed each other in England and Wales: British Assoc. Adv. Sci., Rept. 5th Mtg., Aug. 1835, pp. 59–61

——, 1839, On the older rocks of Devonshire and Cornwall: Geol. Soc. London, Proc., vol. 3, no. 63, pp. 121–123

Sheppard, T., 1917, William Smith: his maps and memoirs: Yorkshire Geol. Soc., Proc., vol. 19, pt. 3, pp. 75–253

Simpson, George Gaylord, 1964, This view of life: the world of an evolutionist: New York, Harcourt, Brace & World

Smith, William, 1806, Observations on the utility, form, and management of water meadows, and the draining and irrigating of peat bogs, with an account of Prisley Bog . . . : Norwich, printed for John Harding

——, 1815-a, A delineation of the strata of England and Wales, with part of Scotland . . . : London, J. Carey

——, 1815-b, A memoir to the map and delineation of the strata of England and Wales, with part of Scotland: London, printed for John Carey

——, 1816–1819, Strata identified by organized fossils, containing prints on colored paper of the most characteristic specimens in each stratum: London, W. Arding

——, 1817, Stratigraphical system of organized fossils, with reference to the specimens of the original geological collection in the British Museum . . . : London, printed for E. Williams

Sollas, William J., 1900, Evolutional geology: Smithsonian Institution, An. Rept., year ending June 30, 1900, pp. 289–314

Sprat, Thomas, 1667, The history of the Royal-Society of London for the improvement of Natural Knowledge: London, J. Martyn and J. Allestry (facsimile reproduction, edited with critical apparatus by J. I. Cope and H. W. Jones: London, Routledge and Kegan Paul, 1959)

Springer, Stewart, 1971, It began with a shark; pp. 309–319 *in* Scherz, Gustav, ed., Dissertations on Steno as geologist: Odense, Odense Univ. Press

Steno, Nicolaus, 1660, Disputatio physica de thermis (A scientific disputation on hot springs), English translation from the Latin text by Alexander J. Pollock, pp. 49–63 *in* Scherz, Gustav, ed., Niels Stensen's geological work: Odense, Odense Univ. Press

——, 1667, Elementorum myologiae specimen, seu musculi descriptio geometrica, cui accedunt Canis carchariae dissectum caput, et Dissectus piscis ex canum genere: Florence, Printing shop under the sign of the star (reproduction of Latin text with English translation by Alexander J. Pollock *in* Scherz, Gustav, ed., Niels Stensen's geological work: Odense, Odense Univ. Press)

——, 1669, Nicolai Stenonis de solido intra solidum naturaliter contento—dissertationis prodromus: Florence, Printing shop under the sign of the star

——, 1669, Discours sur l'anatomie du cerveau: Paris, Robert de Ninville

——, 1671, The prodromus to a dissertation concerning solids naturally contained within solids (English translation by Henry Oldenburg); London, Moses Pitt

Strutt, Robert J. (Fourth Baron Rayleigh), 1905, On the radioactive minerals: Royal Soc. London, Proc., ser. A. vol. 76, pp. 88–101

Tait, Peter Guthrie, 1869, Geological time: North British Rev., vol. 50, pp. 215–233

Takeuchi, H., S. Uyeda, and H. Kanamori, 1970, Debate about the earth: Approach to geophysics through analysis of continental drift: San Francisco, Freeman, Cooper & Co.

Thompson, Silvanus P., 1910, The life of William Thomson, Baron Kelvin of Largs: London, Macmillan, 2 vols.

Tomkeieff, S. I., 1949, James Hutton and the philosophy of geology: Royal Soc. Edinburgh, Proc., sec. B, vol. 63, pp. 387–400

Upham, Warren, 1893, Estimates of geological time: Am. Jour. Sci., 3d. ser., vol. 145, pp. 209–220

Vitaliano, Dorothy B., 1973, Legends of the earth, their geologic origins: Bloomington, Indiana Univ. Press

Wager, Lawrence Rickard, 1964, The history of attempts to establish a quantitative time scale: Geol. Soc. London, Quart. Jour., vol. 120–s, pp. 13–28

Waller, Richard, 1705, The life of Dr. Robert Hooke, pp. i–xxviii *in* The posthumous works of Dr. Robert Hooke (Richard Waller, ed.): London, Smith and Walford. Reprint edition, 1978, by Arno Press, Inc., New York

Warren, Erasmus, 1690, Geologia: or a discourse concerning the earth before the deluge: London, printed for R. Chiswell

Werner, Abraham Gottlob, 1774, On the external characters of minerals (Albert V. Carozzi, tr.): Urbana, Univ. of Illinois Press, 1962

——, 1786, Short classification and description of the various rocks (Alexander Ospovat, tr.): New York, Hafner, 1971

(Whewell, William), 1831, (review of) Principles of geology, . . . by Charles Lyell: British Critic, Quart. Theol. Rev. and Ecclesiastical Record, vol. 9, pp. 180–206

Whiston, William, 1696, A new theory of the earth, from its original, to the consummation of all things: London, R. Roberts

White, George W., 1977, William Maclure's maps of the geology of the United States: Soc. Bibliography Nat. Hist., vol. 8, pp. 266–269

Wilson, Leonard G., 1972, Charles Lyell, the years to 1841: The revolution in geology: New Haven and London, Yale Univ. Press

Winter, John Garrett, 1916, The prodromus of Nicolaus Steno's dissertation concerning a solid body enclosed by process of nature within a solid; an English version with introduction and explanatory notes (foreword by William H. Hobbs): Ann Arbor, Mich., Michigan Humanistic Studies, vol. 11, pt. 2 (facsimile edition edited by George W. White issued in 1968 by Hafner Publishing Co.)

Woodward, Horace B., 1907, The history of the Geological Society of London: London, Geological Soc.

——, 1911, History of geology: New York and London, G. P. Putnam's Sons

Woodward, John, 1695, An essay toward a natural history of the earth: and terrestrial bodies, especially minerals; as also of the sea, rivers, and springs. With an account of the universal deluge: and of the effects that it had upon the earth: London, Ric. Wilkin

Young, G. F., 1923, The Medici: New York, E. P. Dutton and Co., 2 vols.

Zammit-Maempel, George, 1966, St. Paul's tongues and Maltese folklore: Antiquity, vol. 40, no. 159, p. 220

INDEX OF NAMES

SUBJECT INDEX